LONDON MATHEMATICAL SOCIETY STUDENT TEXTS



1 Introduction to combinators and λ-calculus, J.R. HINDLEY & J.P. SELDIN
2 Building models by games, WILFRID HODGES
3 Local fields, J.W.S. CASSELS
4 An introduction to twistor theory, S.A. HUGGETT & K.P. TOD
5 Introduction to general relativity, L. HUGHSTON & K.P. TOD
6 Lectures on stochastic analysis: diffusion theory, DANIEL W. STROOCK
7 The theory of evolution and dynamical systems, J. HOFBAUER & K. SIGMUND
8 Summing and nuclear norms in Banach space theory, G.J.O. JAMESON
9 Automorphisms of surfaces after Nielsen and Thurston, A.CASSON & S. BLEILER
10 Nonstandard analysis and its applications, N.CUTLAND (ed)
11 Spacetime and singularities, G. NABER
12 Undergraduate algebraic geometry, MILES REID
13 An introduction to Hankel operators, J.R. PARTINGTON
14 Combinatorial group theory: a topological approach, DANIEL E. COHEN
16 An introduction to noncommutative noetherian rings, K.R.GOODEARL & R.B.WARFIELD,JR.
17 Aspects of quantum field theory in curved spacetime, S.A. FULLING
18 Braids and coverings: selected topics, VAGN LUNDSGAARD HANSEN

London Mathematical Society Student Texts 18

# Braids and Coverings: selected topics

VAGN LUNDSGAARD HANSEN
Mathematical Institute, Technical University of Denmark

*The right of the University of Cambridge to print and sell all manner of books was granted by Henry VIII in 1534. The University has printed and published continuously since 1584.*

CAMBRIDGE UNIVERSITY PRESS
Cambridge
New York Port Chester Melbourne Sydney

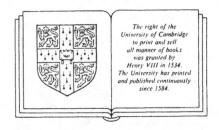

CAMBRIDGE UNIVERSITY PRESS
Cambridge, New York, Melbourne, Madrid, Cape Town,
Singapore, São Paulo, Delhi, Tokyo, Mexico City

Cambridge University Press
The Edinburgh Building, Cambridge CB2 8RU, UK

Published in the United States of America by Cambridge University Press, New York

www.cambridge.org
Information on this title: www.cambridge.org/9780521384797



First published 1989

*A catalogue record for this publication is available from the British Library*

ISBN 978-0-521-38479-7 Hardback
ISBN 978-0-521-38757-6 Paperback

To Birthe, Hanne, Helle and Martin.

# PREFACE

The theories of braids and covering maps offer many suitable topics for a second course in topology accessible to beginning graduate students and still leading to current research. This book is an outgrowth of such a topics course given at the University of Maryland, College Park, in the fall term 1986.

The book has four chapters. The choice of material in the first two chapters, on braid groups with applications to the study of links, has been strongly influenced by the book of Joan Birman: Braids, Links and Mapping Class Groups. It will be apparent that I am much indebted to that book. The material in the final two chapters is, to a large extent, based on my own research on polynomial covering maps. From an algebraic point of view, an n–fold polynomial covering map can be viewed as a homomorphism of the fundamental group of the base space into the Artin braid group on $n$ strings. This is the link between the two themes in the book.

Prerequisites for the book include basic courses in topology, in particular the fundamental group and covering maps, algebra and analysis. It is possible to fill in missing prerequisites along the way, since the concepts treated are very basic and provide good introductions to the relevant subjects. Results from algebraic topology not directly referenced can all be found in the books of Spanier: Algebraic Topology, or Steenrod: The Topology of Fibre Bundles.

It is a pleasure to thank Lars Gæde not only for many discussions during the course of writing but also for contributing Appendix 1 containing a presentation for the coloured (pure) braid group. I am likewise indepted to Hugh Morton for allowing me to reprint his paper "Threading knot diagrams", Math. Proc. Camb. Phil. Soc. 99(1986), 247–260, in Appendix 2. Also thanks to Jesper Michael Møller who read the entire manuscript and suggested many improvements.

Helle Wolter performed a skilful and efficient typing job and Beth Beyerholm made most af the figures.

Finally, I want to extend my gratitude to the students in my class at College Park, who made it such a delight to present these lectures.

Lyngby, August 1989 Vagn Lundsgaard Hansen.

# CONTENTS

## Chapter III. POLYNOMIAL COVERING MAPS.



## Chapter IV. ALGEBRA AND TOPOLOGY OF WEIERSTRASS POLYNOMIALS.

Chapter I

# BRAIDS AND CONFIGURATION SPACES

Braids are among the oldest inventions of mankind. They are used for practical purposes to make rope, and for decorations in weaving patterns and hairstyle etc.

As mathematical objects they were introduced by the German mathematician Emil Artin in a paper from 1925, although the idea was implicit in a paper of Hurwitz from 1891. Artin proposed to use braids to study knots and links. There are now applications of braids in several mathematical subjects: Topology (knots, links, fixed point theory, covering spaces), geometry (mapping class groups), singularity theory, dynamical systems, operator algebras, to mention a few. Such developments and many more were reported in a meeting on Artin's braid group held in Santa Cruz, July 13 to July 26, 1986, and has appeared in the conference proceedings (edited by J.S. Birman and A. Libgober) as volume 78 in the AMS Contemporary Mathematics series.

Chapter I contains the basic material on braid groups and the spaces related to them. The chapter opens with the definition of a geometric braid on $n$ strings, first the original definition of Artin from 1925 as a system of $n$ strings between two parallel planes in euclidean 3–space, and then the equivalent definition – suggested by Fox 1962 – as a loop in the space of configurations of a set of $n$ points in the euclidean plane. The definition of Fox can be extended to define braids in any manifold and will prevail in the book. The fundamental system of fibrations of configuration spaces defined by Fadell and Neuwith 1962 plays an important role. We prove the presentation theorem of the Artin braid group on $n$ strings in terms of generators and relations, and the representation theorem of this group as a subgroup of the group of automorphisms of the free group on $n$ generators. In the final section of Chapter I we use braids in the 2–sphere to solve the Dirac string problem. It appears to be the first time this application of braid theory is included in a book.

## 1. Geometric braids.

Let $\mathbb{E}^3$ denote euclidean 3–space. We identify $\mathbb{E}^3$ with the 3–dimensional real number space $\mathbb{R}^3$ by choosing a coordinate system with coordinates $(x,y,z)$, in which the Z–axis points vertically downwards. See Figure 1.

Consider two horizontal parallel planes in $\mathbb{E}^3$ with constant z–coordinates $z_0$ and $z_1$ respectively, where $z_0 < z_1$. We call the plane $z = z_0$ the upper plane and the plane $z = z_1$ the lower plane. Mark n different points $P_1,\dots,P_n$ on a line in the upper plane and project them orthogonally onto the lower plane to the points $P_1',\dots,P_n'$.

Definition 1.1. A geometric braid on n strings (or an n–string braid, or just an n–braid) $\beta$ is a system of n embedded arcs $\mathcal{A} = \{\mathcal{A}_1,\dots,\mathcal{A}_n\}$ in $\mathbb{E}^3$, where the $i$th arc $\mathcal{A}_i$ connects the point $P_i$ on the upper plane to the point $P'_{\tau(i)}$ on the lower plane for some permutation $\tau$ of $\{1,\dots,n\}$, such that

(i) Each arc $\mathcal{A}_i$ intersects each intermediate parallel plane between the upper and the lower plane exactly once.

(ii) The arcs $\mathcal{A}_1,\dots,\mathcal{A}_n$ intersect each intermediate parallel plane between the upper and the lower plane in exactly n different points.

The permutation $\tau$ is called the permutation of the braid. The arc $\mathcal{A}_i$ is called the $i$th string (or strand) in the braid.

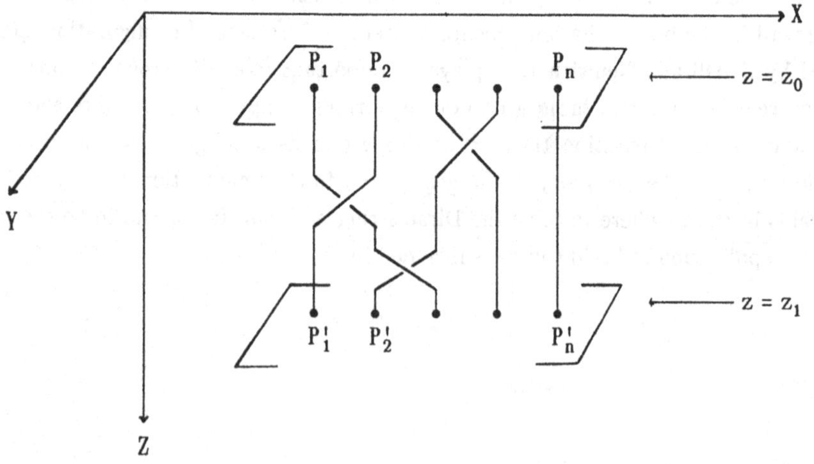


Figure 1

We think of an arc in $\mathbb{E}^3$ as the image of an embedding $\mathscr{A}_i\colon [0,1] \to \mathbb{E}^3$ of the unit interval $[0,1]$ into $\mathbb{E}^3$. We use the same notation for the arc and the corresponding embedding. As indicated, we think of a braid as hanging downwards.

To make braids into a useful concept we need to define equivalence of braids.

Definition 1.2. Two n–braids $\mathscr{A}^0 = \{\mathscr{A}_1^0,\dots,\mathscr{A}_n^0\}$ and $\mathscr{A}^1 = \{\mathscr{A}_n^1,\dots,\mathscr{A}_n^1\}$ with the same permutation $\tau$, are called equivalent (or homotopic), if there is a homotopy through geometric braids with permutation $\tau$ from $\mathscr{A}^0$ to $\mathscr{A}^1$, in other words, if there exist n continuous maps

$$F_i\colon [0,1] \times [0,1] \to \mathbb{E}^3 \ , \quad 1 \le i \le n \ ,$$

such that

$$\left.\begin{aligned} F_i(t,0) &= \mathscr{A}_i^0(t) \\ F_i(t,1) &= \mathscr{A}_i^1(t) \end{aligned}\right\} 0 \le t \le 1 \ , \quad 1 \le i \le n$$

$$\left.\begin{aligned} F_i(0,s) &= P_i \\ F_i(1,s) &= P'_{\tau(i)} \end{aligned}\right\} 0 \le s \le 1 \ , \quad 1 \le i \le n$$

and such that if we define $\mathscr{A}_i^s\colon [0,1] \to \mathbb{E}^3$ by $\mathscr{A}_i^s(t) = F_i(t,s)$, then $\mathscr{A}^s = \{\mathscr{A}_1^s,\dots,\mathscr{A}_n^s\}$ is a geometric n–braid (with permutation $\tau$) for each $0 \le s \le 1$.

Remark 1.3. There are several other possible notions of equivalence of geometric n–braids, e.g. ambient isotopy in $\mathbb{E}^3$ (through geometric n–braids) keeping the regions in $\mathbb{E}^3$ corresponding to $z \le z_0$ and $z \ge z_1$ pointwise fixed. As proved by Artin in 1947, [23], this notion leads to the same equivalence classes.

Throughout the remainder of the book, we shall not distinguish in notation between the equivalence class of a braid and the braid itself.

After a slight homotopy, we can (and will) assume that a braid $\beta$ consists of polygonal arcs only, and that we get transversal crossings of the arcs if we project the braid orthogonally onto the plane in $\mathbb{E}^3$ containing the points $P_1,\dots,P_n,P'_1,\dots,P'_n$. By this projection we get a standard picture of the braid $\beta$ as shown in Figure 2. Also note that (up to equivalence) we may assume that the crossings of strings occur on different levels, and that over and under crossings of strings must be indicated.

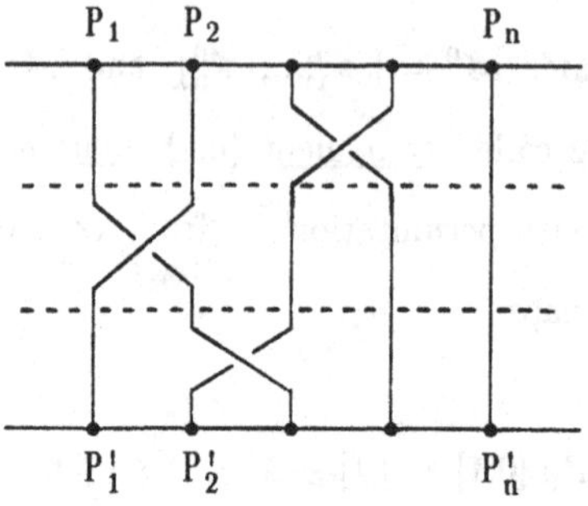


Figure 2

In Figure 2, we have indicated that a braid can be resolved into elementary braids, in which all strings except a neighbouring pair of strings go straight from the top to the bottom and the neighbouring pair just interchange.

For $1 \leq i \leq n-1$, we denote by $\sigma_i$ that elementary geometric n–braid, in which the *i*th string just overcrosses the *(i+1)*th string once and all other strings go straight from the top to the bottom. See Figure 3.

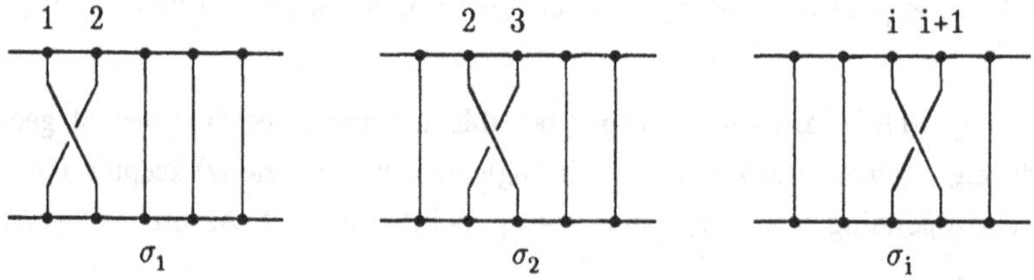


Figure 3

Let B(n) denote the set of all equivalence classes of geometric n–braids. It turns out that this set can be equipped with a natural group structure, which we shall now define.

Let $\beta_1$ and $\beta_2$ be geometric n–braids . Then we define the product (composition) of $\beta_1$ and $\beta_2$ , denoted $\beta_1 \cdot \beta_2$ , as follows: First hang the braid $\beta_2$ under the braid $\beta_1$ by attaching the lower plane of $\beta_1$ to the upper plane of $\beta_2$ . Then remove the plane along which the braids $\beta_1$ and $\beta_2$ are attached to each other. Now squeeze the resulting system of arcs (strings) to lie between the planes $z = z_0$ and $z = z_1$ , and we have the braid $\beta_1 \cdot \beta_2$ . See Figure 4.

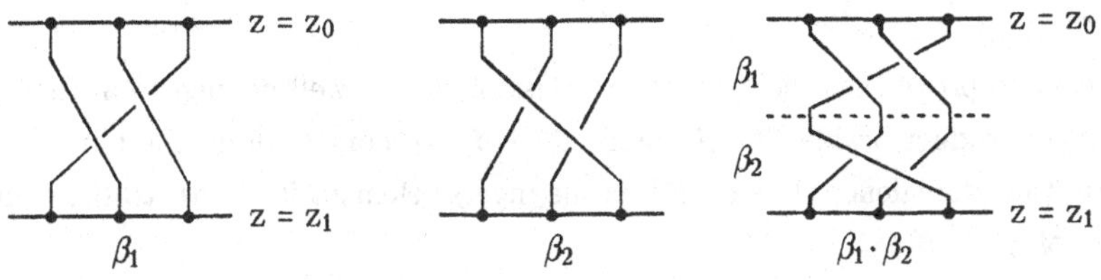


Figure 4

If we substitute homotopic braids $\beta_1'$ and $\beta_2'$ for $\beta_1$ and $\beta_2$ respectively, then it easy to prove that the product braids $\beta_1 \cdot \beta_2$ and $\beta_1' \cdot \beta_2'$ are homotopic. Thus the product is well defined on equivalence classes of n–braids and induces a product in B(n) .

The trivial n–braid $\epsilon$ is the n–braid in which all strings just go straight from the upper plane to the lower plane. The projection of $\epsilon$ is shown in Figure 5. It is easily seen that the equivalence class of $\epsilon$ is a neutral element for the product in B(n) .

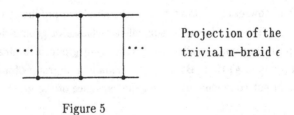

Projection of the trivial n–braid $\epsilon$

Figure 5

The inverse braid $\beta^{-1}$ of the braid $\beta$ is defined as the mirror image of $\beta$ with respect to a horizontal plane between the upper plane and the lower plane. The projections of $\beta$ and $\beta^{-1}$ are shown in Figure 6.

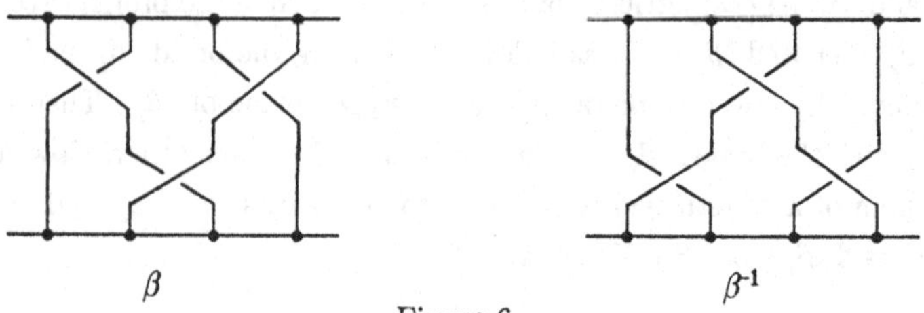


Figure 6

It is easy to prove that the equivalence class of $\beta^{-1}$ is well defined from that of $\beta$ and that the product braids $\beta \cdot \beta^{-1}$ and $\beta^{-1} \cdot \beta$ are homotopic to the trivial braid. Therefore, the equivalence class of $\beta^{-1}$ is the inverse element in B(n) to the equivalence class of $\beta$.

For the elementary n–braid $\sigma_i$, $1 \leq i \leq n-1$, the inverse braid $\sigma_i^{-1}$ is obtained by changing (in the standard projection) the overcrossing of the i + 1 string by the i string to an undercrossing. See Figure 7.

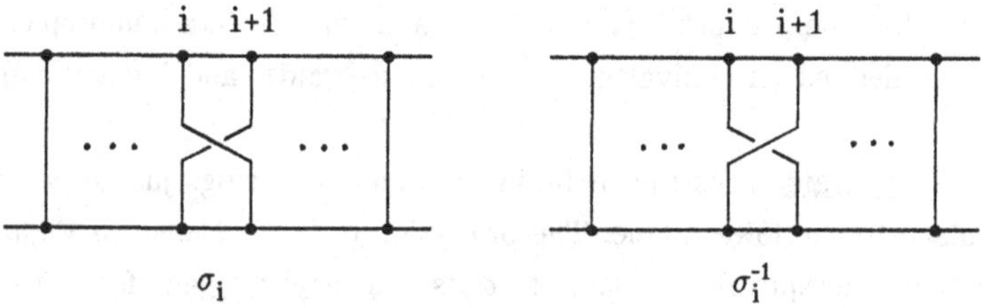


Figure 7

It is now easy to prove that with the above product, neutral element and inverse elements, the set B(n) of equivalence classes of geometric n–braids is actually a group. This group is called the Artin braid group of braids on n strings. Since we shall prove in §3 that B(n) is the fundamental group of a suitable topological space, we shall not comment further on the existence of the group structure in B(n) at this point.

Remark 1.4. The 5–braid $\beta$ with projection as in Figure 2 – or, more precisely, its equivalence class – can be written as the product $\beta = \sigma_3^{-1} \cdot \sigma_1^{-1} \cdot \sigma_2$. Notice that the elementary braid $\sigma_i$ may occur in n–braids for all $n \geq i + 1$ . Hence it is important to know the number of strings in a braid $\beta$ written as a product of elementary braids and their inverses.

As already indicated in Figure 2, it is intuitively clear that the equivalence class of any n–braid can be written as a product of the elementary n–braids $\sigma_i$ , $1 \leq i \leq n-1$ , and their inverses. In other words: The elementary n–braids $\sigma_1,\ldots,\sigma_{n-1}$ generate the group B(n) .

We shall now look for relations among the elements in B(n) . First we notice that if $|i-j| \geq 2$ and $1 \leq i,j \leq n-1$ , then – since the pair consisting of the i and i + 1 string does not interfere with the pair consisting of the j and j + 1 string – we get the following relation

$$(1) \quad \sigma_i \cdot \sigma_j = \sigma_j \cdot \sigma_i \quad \text{for} \quad |i-j| \geq 2 \, , 1 \leq i \, , j \leq n-1 \ .$$

Figure 8 illustrates relation (1) .

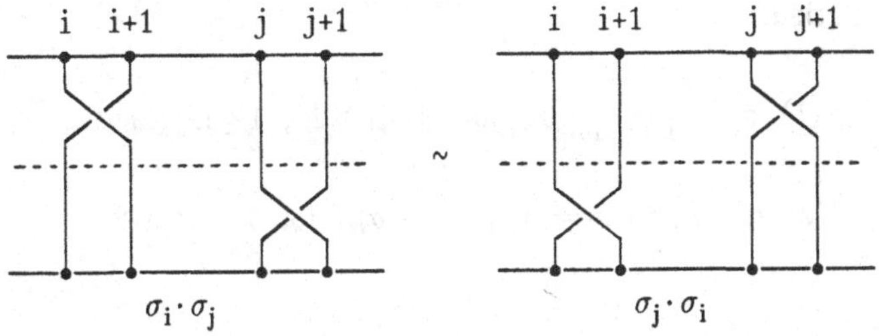

Figure 8

As illustrated in Figure 9, we also have the following relation in B(n)

$$(2) \quad \sigma_i \cdot \sigma_{i+1} \cdot \sigma_i = \sigma_{i+1} \cdot \sigma_i \cdot \sigma_{i+1} \quad \text{for} \ 1 \leq i \leq n-2 \ .$$

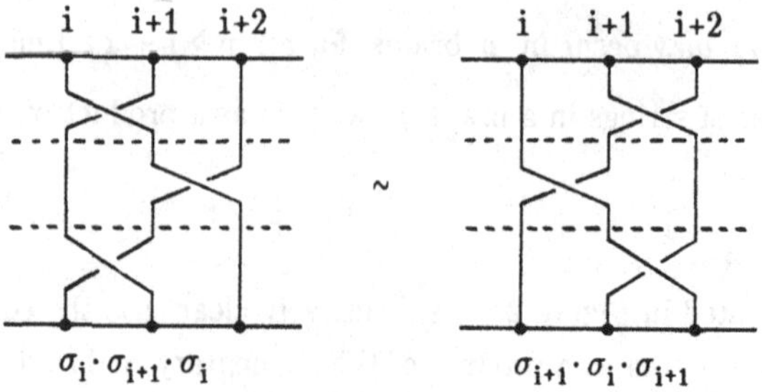


Figure 9

Already in Artin's first paper on braid groups [22] from 1925, a proof was given that the relations (1) and (2) generate all relations among the elements in B(n). This is actually not quite trivial. We state the result below as Theorem 1.5, but postpone the proof until §4.

Theorem 1.5. The group B(n) of geometric braids on n strings admits a presentation with

generators: $\sigma_1, \sigma_2, \ldots, \sigma_{n-1}$

and defining relations:

(1) $\sigma_i \cdot \sigma_j = \sigma_j \cdot \sigma_i$ for $|i-j| \geq 2$ , $1 \leq i, j \leq n-1$

(2) $\sigma_i \cdot \sigma_{i+1} \cdot \sigma_i = \sigma_{i+1} \cdot \sigma_i \cdot \sigma_{i+1}$ for $1 \leq i \leq n-2$ .

2. Configuration spaces of ordered finite pointsets and their fibrations.

Let $M$ denote a connected manifold of dimension $\geq 2$. For an integer $n \geq 1$, consider the subset $F_n(M)$ in the n-fold product $M\times\ldots\times M$ of $M$ with itself, consisting of all n-tuples of pairwise different points in $M$, i.e.

$$F_n(M) = \{(x_1,\ldots,x_n) \in M\times\ldots\times M \mid x_i \neq x_j \text{ for } i \neq j\} .$$

We can think of $F_n(M)$ as the space of possible configurations for a set of $n$ ordered points in $M$. Accordingly, $F_n(M)$ is called the configuration space for a set of $n$ (ordered) points in $M$. We equip $F_n(M)$ with the topology induced from the product topology on $M\times\ldots\times M$.

The configuration space $F_n(M)$ can be constructed from $M\times\ldots\times M$ by removing the finitely many submanifolds where any two coordinates are equal. Since the codimension of these submanifolds in $M\times\ldots\times M$ equals the dimension of $M$, and hence by assumption is $\geq 2$, a simple transversality argument shows, therefore, that $F_n(M)$ is connected. Hence in particular, the homotopy groups of $F_n(M)$ are independent of base point.

Now let $m \geq 0$ be a nonnegative integer. By $Q_m = \{q_1,\ldots,q_m\}$ we denote a fixed, but arbitrary, subset of $M$ consisting of $m$ pairwise different points $q_i \in M$. For $m = 0$, take $Q_0$ to be the empty set. In order to study the configuration space $F_n(M)$, it is fruitful also to consider the configuration spaces

$$F_{m,n}(M) = F_n(M\backslash Q_m) .$$

If $Q'_m = \{q'_1,\ldots,q'_m\}$ is another subset of $M$ containing $m$ pairwise different points, then it is well known that there exists a homeomorphism on $M$ taking $Q_m$ to $Q'_m$. Hence the complements $M\backslash Q_m$ and $M\backslash Q'_m$ are homeomorphic open sets in $M$, and therefore the spaces $F_n(M\backslash Q_m)$ and $F_n(M\backslash Q'_m)$ are homeomorphic. Consequently, the topological type of the configuration space $F_{m,n}(M)$ is independent of the particular subset $Q_m \subset M$ chosen.

We note that

$$F_{0,n}(M) = F_n(M) \quad \text{and} \quad F_{m,1}(M) = M\backslash Q_m .$$

The following fundamental theorem was proved by Fadell and Neuwirth [30] in 1962.

Theorem 2.1. Suppose $n \geq 2$ and let $1 \leq r < n$ . Then the map

$$\pi : F_{m,n}(M) \to F_{m,r}(M) ,$$

defined by

$$\pi(x_1,\ldots,x_n) = (x_1,\ldots,x_r) ,$$

is a locally trivial fibration with fibre $F_{m+r,n-r}(M)$ .

Proof Consider a point $(x_1^0,\ldots,x_r^0) \in F_{m,r}(M)$ .

First we shall describe the inverse image under $\pi$ of this point, $\pi^{-1}(x_1^0,\ldots,x_r^0)$ , which eventually will be the typical fibre of the fibration $\pi$ . Clearly,

$$\pi^{-1}(x_1^0,\ldots,x_r^0) = \{(x_1^0,\ldots,x_r^0 , y_1,\ldots,y_{n-r}) \mid \text{All coordinates different and in } M\backslash Q_m \} .$$

If we put $Q_{m+r} = Q_m \cup \{x_1^0,\ldots,x_r^0\}$ , then

$$F_{m+r,n-r}(M) = \left\{ (y_1,\ldots,y_{n-r}) \;\middle|\; \begin{matrix} y_i \neq y_j & \text{for } i \neq j \\ y_i \in M\backslash Q_{m+r} & \end{matrix} \right\} ,$$

and there is an obvious homeomorphism

$$h : F_{m+r,n-r}(M) \to \pi^{-1}(x_1^0,\ldots,x_r^0)$$

defined by

$$h(y_1,\ldots,y_{n-r}) = (x_1^0,\ldots,x_r^0 , y_1,\ldots,y_{n-r}) .$$

The proof of local triviality of $\pi$ will be carried out only for $r = 1$ , mainly for notational convenience. For the rest of the proof, we shall therefore only consider the map

$$\pi : F_{m,n}(M) \to F_{m,1}(M) ,$$

defined by $\pi(x_1,\ldots,x_n) = x_1$ .

We shall prove that $\pi$ is trivial over a neighbourhood $U$ of an arbitrary point $x_0 \in F_{m,1}(M) = M\backslash Q_m$ . As above we let $Q_{m+1} = Q_m \cup \{x_0\}$ .

Let $U$ denote a neighbourhood of $x_0$ in the open set $M \backslash Q_m$ in $M$, which is homeomorphic to an open ball, and let $\bar{U}$ denote the closure of $U$. Define a (continuous) map $\Theta : U \times \bar{U} \to \bar{U}$ with the following properties. Setting $\Theta_x(y) = \Theta(x,y)$ we require:

(i) $\Theta_x : \bar{U} \to \bar{U}$ is a homeomorphism, which fixes the boundary $\partial\bar{U}$ of $\bar{U}$ pointwise.

(ii) $\Theta_x(x) = x_0$ .

In Sublemma 2.2 below, we shall prove that such a map $\Theta$ does exist.

By property (i), $\Theta$ can be extended to a (continuous) map $\Theta : U \times M \to M$ by setting $\Theta(x,y) = y$ for $y \notin U$.

The homeomorphism $\Theta_x : M \to M$ takes pairwise different $(n-1)$ – tuples of points in $M$ avoiding $x \in U$ into pairwise different $(n-1)$ – tuples of points in $M$ avoiding $x_0 \in U$. In other words, $\Theta_x$ maps the "fibre" of $\pi$ over $x$ homeomorphically onto the "fibre" of $\pi$ over $x_0$. Then it follows easily that the required local product representation of $\pi$,

$$U \times F_{m+1,n-1}(M) \underset{\Phi^{-1}}{\overset{\Phi}{\rightleftarrows}} \pi^{-1}(U) ,$$

is given by

$$\Phi(x, y_1,\ldots,y_{n-1}) = (x, \Theta_x^{-1}(y_1),\ldots,\Theta_x^{-1}(y_{n-1}))$$

$$\Phi^{-1}(x, y_1,\ldots,y_{n-1}) = (x, \Theta_x(y_1),\ldots,\Theta_x(y_{n-1})) .$$

To finish the proof of Theorem 2.1, it only remains to prove

Sublemma 2.2. There exists a continuous map $\Theta : U \times \bar{U} \to \bar{U}$ with the properties (i) and (ii) as above.

Proof Let $\mathbb{R}^n$ denote the n–dimensional real number space with euclidean norm $\|\cdot\|$. To prove Sublemma 2.2, it is sufficient to consider the unit ball $U$ in $\mathbb{R}^n$ and $x_0 = 0$. Then $\bar{U} = \{y \in \mathbb{R}^n \mid \|y\| \leq 1\}$ .

For $x \in U$, i.e. $\|x\| < 1$, choose a function

$$\lambda_x : \bar{U} \to \mathbb{R}_0^+ ,$$

where $\mathbb{R}_0^+$ denotes the nonnegative real numbers, such that

$$\lambda_x(y) = \begin{cases} 1 & \text{for } 0 \le \|y\| \le \|x\| + \frac{1}{3}(1 - \|x\|) \\ 0 & \text{for } \|x\| + \frac{2}{3}(1 - \|x\|) \le \|y\| \le 1 , \end{cases}$$

and such that $\lambda : U \times \bar{U} \to \mathbb{R}_0^+$ defined by $\lambda(x,y) = \lambda_x(y)$ is continuously differentiable.

For $x \in U$ define now a vector field $v_x$ on $\bar{U}$ by

$$v_x(y) = \lambda_x(y) \cdot (0 - x) ,$$

and let $\theta_x^t(y)$, $t \in \mathbb{R}$, be the corresponding flow. Then the homeomorphism $\theta_x(y) = \theta_x^1(y)$ is as desired. □

We shall now consider the special case, where we take the manifold M to be the euclidean plane $\mathbb{E}^2$.

First note that the space $\mathbb{E}^2 \backslash Q_m$ for $m \ge 1$ has the homotopy type of a wedge (bouquet) of m copies of the 1-sphere $S^1$. See Figure 10.

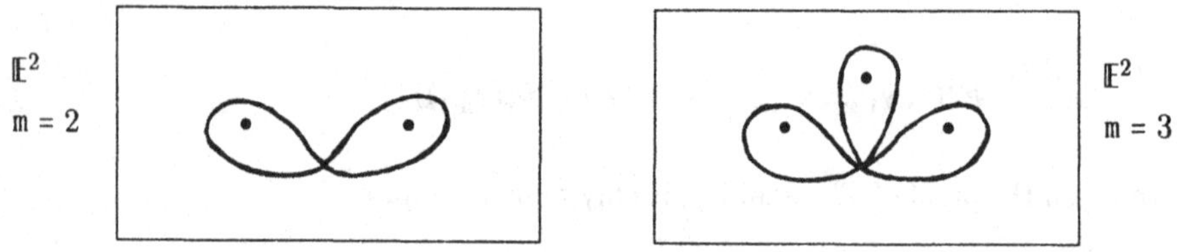


Figure 10

In particular we note that all homotopy groups of $\mathbb{E}^2 \backslash Q_m$ in dimensions $i \ge 2$ vanish. From Theorem 2.1, we therefore deduce the following

<u>Corollary 2.3</u> (Fadell and Neuwirth). For the configuration space $F_n(\mathbb{E}^2)$ of n ordered points in the plane $\mathbb{E}^2$, we have

$$\pi_i(F_n(\mathbb{E}^2)) = 0 \text{ for } i \ge 2 .$$

More generally: For all $n \geq 1$ and $m \geq 0$ we have

$$\pi_i(F_{m,n}(\mathbb{E}^2)) = 0 \text{ for } i \geq 2 \text{ .}$$

<u>Proof</u> For notational simplicity, put $F_{m,n} = F_{m,n}(\mathbb{E}^2)$ , and consider the following diagram

$$\begin{array}{ccccccccccc}
F_{n-1,1} & \rightarrow & F_{n-2,2} & \rightarrow & \cdots & \rightarrow & F_{2,n-2} & \rightarrow & F_{1,n-1} & \rightarrow & F_{0,n} = F_n(\mathbb{E}^2) \\
 & & \downarrow & & & & \downarrow & & \downarrow & & \downarrow \\
 & & F_{n-2,1} & & & & F_{2,1} & & F_{1,1} & & F_{0,1} \quad ,
\end{array}$$

in which the vertical maps are the fibrations of Theorem 2.1, and the horizontal maps are the inclusions of the fibres in the corresponding total spaces of the fibrations.

For the spaces $F_{m,1}(\mathbb{E}^2) = \mathbb{E}^2 \backslash Q_m$ , all homotopy groups in dimensions $i \geq 2$ vanish. Working through the homotopy sequences for the fibrations, from the left to the right in the diagram, we conclude in the final stage that

$$\pi_i(F_n(\mathbb{E}^2)) = \pi_i(F_{0,n}) = 0 \text{ for } i \geq 2 \text{ .}$$

More generally, as an intermediate step towards proving that $\pi_i(F_{0,n+m}) = 0$ , we prove that

$$\pi_i(F_{m,n}(\mathbb{E}^2)) = \pi_i(F_{m,n}) = 0 \text{ for } i \geq 2 \text{ .} \qquad \square$$

## 3. The braid group as fundamental group of a configuration space.

As in §2, let M denote a connected manifold of dimension $\geq 2$.

Let $\Sigma_n$ denote the symmetric group on n elements, i.e. the group of all permutations of the set $\{1,...,n\}$.

There is a natural right action of $\Sigma_n$ on the configuration space $F_n(M)$,

$$\mu : F_n(M) \times \Sigma_n \to F_n(M) ,$$

defined by permutation of coordinates, i.e.

$$\mu((x_1,...,x_n),\sigma) = (x_1,...,x_n) \cdot \sigma = (x_{\sigma(1)},...,x_{\sigma(n)}) .$$

It is easy to prove that $\mu$ is indeed a right action. Since all coordinates for points in $F_n(M)$ are pairwise different, it is, moreover, a free action.

Denote the orbit space for the free action $\mu$ by $C_n(M)$, or in standard notation,

$$C_n(M) = {F_n(M)}/{\Sigma_n} .$$

The orbits of elements in $F_n(M)$ under the action $\mu$ consist of n–tuples $(x_1,..,x_n)$ of pairwise different points $x_i \in M$, where two n–tuples are on the same orbit if they differ only by a permutation. Hence, one can think of $C_n(M)$ as the space of all possible configurations for an unordered set of n pairwise different points in M. Accordingly, $C_n(M)$ is called the configuration space for a set of n (unordered) points in M.

Projection of $F_n(M)$ onto the orbit space $C_n(M)$ defines a principal $\Sigma_n$–bundle,

$$p_n : F_n(M) \to C_n(M) .$$

In particular, $p_n$ is an n!–fold covering map.

Following an idea of Fox [31], configuration spaces can be used to define a notion of braids in M, which specializes to the geometric braids defined by Artin in the case $M = \mathbb{E}^2$. This will now be explored.

Choose a base point $\overline{c}_0$ in $F_n(M)$, and take $c_0 = p_n(\overline{c}_0)$ as base point in $C_n(M)$. Consider an arbitrary element $\beta \in \pi_1(C_n(M),c_0)$ in the fundamental group of $C_n(M)$, represented by the closed loop $f: [0,1] \to C_n(M)$ with $f(0) = f(1) = c_0$. There is a unique lifting of f in the covering map $p_n : F_n(M) \to C_n(M)$ to a path

$\bar{f}: [0,1] \to F_n(M)$ in $F_n(M)$ with $\bar{f}(0) = \bar{c}_0$. One can think of $\bar{f} = (\bar{f}_1,...,\bar{f}_n)$ as n paths $\bar{f}_i: [0,1] \to M$, $1 \leq i \leq n$, in M, for which $\bar{f}_i(t) \neq \bar{f}_j(t)$ for all $i \neq j$ and all $t \in [0,1]$. Note that the ordered set of points $(\bar{f}_1(1),...,\bar{f}_n(1))$ in M is just a permutation $\tau$ of the ordered set of points $(\bar{f}_1(0),...,\bar{f}_n(0))$ in M, since $p_n(\bar{f}(1)) = p_n(\bar{f}(0)) = c_0$.

Define embeddings $\mathcal{A}_i: [0,1] \to M \times [0,1]$, $1 \leq i \leq n$, by $\mathcal{A}_i(t) = (\bar{f}_i(t), t)$. The system $\mathcal{A} = \{\mathcal{A}_1,..,\mathcal{A}_n\}$ is then a system of n strings in the product space $M \times [0,1]$ satisfying the same formal requirements as laid down in Definition 1.1 for a geometric braid. In particular, the system of strings $\mathcal{A}$ in $M \times [0,1]$ connects the set of points $P_1,...,P_n$ in $M = M \times \{0\}$, which is the set of coordinates for $\bar{c}_0 \in F_n(M)$, to the corresponding set of points $P'_1,...,P'_n$ in $M \times \{1\}$, according to the permutation $\tau$ that turns $\bar{f}(0)$ into $\bar{f}(1)$.

The homotopy class of $\mathcal{A}$ defined in analogy with Definition 1.2 is well defined since it only depends on $\beta$ and not on the particular representative $f: [0,1] \to C_n(M)$ originally chosen.

Following Fox, we call $\mathcal{A}$ – or, for the sake of simplicity, the original homotopy class $\beta \in \pi_1(C_n(M),c_0)$ itself – a <u>braid with permutation $\tau$ in M on n strings</u>, and the group $\pi_1(C_n(M),c_0)$ <u>the group of braids on n strings in M.</u>

The group $\pi_1(F_n(M),\bar{c}_0)$ is called the Fox group of <u>coloured</u> (or <u>pure</u>) braids on n strings in M. Coloured braids are exactly the braids with trivial permutation of the strings.

Now we specialize and take the manifold M to be the euclidean plane $\mathbb{E}^2$. We shall consider $\mathbb{E}^2$ as embedded in euclidean space $\mathbb{E}^3$, such that after identification of $\mathbb{E}^3$ with $\mathbb{R}^3$, the plane $\mathbb{E}^2$ corresponds to $\mathbb{R}^2 = \mathbb{R}^2 \times \{0\}$.

Choose points

$$P_1 = (1,0,0),\ P_2 = (2,0,0),....,P_n = (n,0,0) \quad \text{in the plane } z = 0$$

and
$$P'_1 = (1,0,1),\ P'_2 = (2,0,1),....,P'_n = (n,0,1) \quad \text{in the plane } z = 1\,.$$

Let

$$p_n: F_n(\mathbb{E}^2) \to C_n(\mathbb{E}^2)$$

be the canonical pricipal $\Sigma_n$ – bundle of configuration spaces, and take $\bar{c}_0 = (P_1,...,P_n) \in F_n(\mathbb{E}^2)$ as base point in $F_n(\mathbb{E}^2)$ and $c_0 = p_n(\bar{c}_0) \in C_n(\mathbb{E}^2)$ as the corresponding base point in $C_n(\mathbb{E}^2)$.

Then the braid $\mathscr{B}$ in $\mathbb{E}^2$ in the sense of Fox corresponding to a homotopy class $\beta \in \pi_1(C_n(\mathbb{E}^2),c_0)$ as defined above is clearly a geometric braid on n strings in the sense of Artin as defined in Definition 1.1. Conversely, it is also clear how to define a Fox braid from an Artin braid.

It is easy to prove that the product of equivalence classes of geometric n–braids defined in §1 corresponds to the standard product of homotopy classes of loops in $C_n(\mathbb{E}^2)$. Also formation of inverses in the two algebraic structures correspond to each other. Hence we get

Theorem 3.1. The Artin braid group B(n) can be canonically identified with the fundamental group $\pi_1(C_n(\mathbb{E}^2),c_0)$.

In particular, we now have a full proof that B(n) is actually a group. In the following, we shall freely identify B(n) and $\pi_1(C_n(\mathbb{E}^2),c_0)$ when convenient.

The homotopy sequence for the principal $\Sigma_n$ – bundle (n!–fold covering map) $p_n$: $F_n(\mathbb{E}^2) \to C_n(\mathbb{E}^2)$ reduces by Corollary 2.3 to the following short exact sequence:

$$1 \to \pi_1(F_n(\mathbb{E}^2),\bar{c}_0) \xrightarrow{\rho_n} \pi_1(C_n(\mathbb{E}^2),c_0) \xrightarrow{\tau_n} \Sigma_n \to 1 .$$

If we use the identification

$$B(n) = \pi_1(C_n(\mathbb{E}^2),c_0) ,$$

and put

$$H(n) = \pi_1(F_n(\mathbb{E}^2),\bar{c}_0) ,$$

then the elements in H(n) can be identified with the (equivalence classes of) geometric braids on n strings with trivial permutation of the strings. As mentioned earlier, these braids are called coloured, or pure, braids.

The boundary homomorphism $\tau_n$ in the above short exact sequence is defined by lifting in the covering map $p_n$, and hence $\tau_n(\beta)$ is the permutation of the braid $\beta$ for every braid $\beta \in B(n)$. The homomorphism $\rho_n$ is induced by the map $p_n$.

Introducing these identifications in the above short exact sequence, we get the short exact sequence

$$1 \longrightarrow H(n) \xrightarrow{\rho_n} B(n) \xrightarrow{\tau_n} \Sigma_n \longrightarrow 1 \ ,$$

which is called the braid group sequence. For obvious reasons, the homomorphism $\tau_n$ is called the permutation homomorphism.

From Corollary 2.3 we get also the following

Theorem 3.2. The configuration space $C_n(\mathbb{E}^2)$ is an Eilenberg – MacLane space of type $(B(n),1)$, i.e. $\pi_1(C_n(\mathbb{E}^2)) = B(n)$ and $\pi_i(C_n(\mathbb{E}^2)) = 0$ for $i \geq 2$.

If G is the fundamental group of a finite dimensional CW–complex K, which is an Eilenberg – MacLane space of type (G,1), then a theorem of P.A. Smith implies that G contains no nontrivial elements of finite order. [A short proof: Suppose G contains an element of order m. Then K admits a covering space $\tilde{K}$, which is again a finite dimensional Eilenberg – MacLane space, with fundamental group cyclic of order m. Obtain a contradiction with ([7], Theorem IV.7.1 and Theorem IV.11.5).] Since $C_n(\mathbb{E}^2)$ is a finite dimensional Eilenberg – MacLane space of type $(B(n),1)$, we get then the following

Corollary 3.3. The braid group $B(n)$ contains no nontrivial elements of finite order.

Corollary 3.3 was proved first by Fox, Fadell and Neuwirth in 1962. Only recently in 1980 has a purely group theoretical proof of this result been given by Joan Dyer [27].

## 4. Artin's presentation of the braid group.

In §1 we have seen that the Artin braid group on n strings B(n) is generated by the elementary braids $\sigma_1,...,\sigma_{n-1}$, where $\sigma_i$ just interchanges the i and i + 1 strings. Also, two simple relations among these element were established. Our aim in this section is to prove that this is actually a presentation of B(n).

We shall use the identification

$$B(n) = \pi_1(C_n(\mathbb{E}^2),c_0)$$

described in §3.

In accordance with §3, we use the set of points $P_1 = (1,0), P_2 = (2,0),...,P_n = (n,0)$ as the set of initial points for the strings in the n–braids. Then the elementary n–braid $\sigma_i$, $1 \leq i \leq n-1$, can be described by the path $\bar{f}: [0,1] \to F_n(\mathbb{E}^2)$, defined by

$$\bar{f}(t) = (P_1,...,P_{i-1},\bar{f}_i(t),\bar{f}_{i+1}(t),P_{i+2},...,P_n),$$

where $\bar{f}_i(t) = (i + \sin\frac{\pi}{2}t, \sin \pi t)$ and $\bar{f}_{i+1}(t) = (i + \cos\frac{\pi}{2}t, -\sin \pi t)$.
Since

$$\bar{f}(0) = (P_1,...,P_i,P_{i+1},...,P_n)$$

and

$$\bar{f}(1) = (P_1,...,P_{i+1},P_i,...,P_n),$$

the path $\bar{f}: [0,1] \to F_n(\mathbb{E}^2)$ projects to a loop $f: [0,1] \to C_n(\mathbb{E}^2)$ with $f(0) = f(1) = c_0$, and this loop represents $\sigma_i$ in $\pi_1(C_n(\mathbb{E}^2),c_0)$.

The following theorem, which was stated as Theorem 1.5, is our objective. It is in Artin's paper [22] from 1925, but we present an alternative proof due to Fadell and van Buskirk [29].

<u>Theorem</u> (Artin's presentation of the braid group). The group B(n) of geometric braids on n strings admits a presentation with

generators: $\sigma_1,...,\sigma_{n-1}$

and defining relations:

$$(1)\ \sigma_i \cdot \sigma_j = \sigma_j \cdot \sigma_i \ \text{ for } \ |i-j| \geq 2\ ,\ 1 \leq i, j \leq n-1\ .$$

$$(2)\ \sigma_i \cdot \sigma_{i+1} \cdot \sigma_i = \sigma_{i+1} \cdot \sigma_i \cdot \sigma_{i+1} \ \text{ for } \ 1 \leq i \leq n-2\ .$$

Proof (due to Fadell and van Buskirk).

Let $B_n$ denote the abstract group with presentation given by

$$\text{generators: } \tilde{\sigma}_1, \ldots, \tilde{\sigma}_{n-1}$$

and defining relations:

$$(1)\ \tilde{\sigma}_i \cdot \tilde{\sigma}_j = \tilde{\sigma}_j \cdot \tilde{\sigma}_i \ \text{ for } \ |i-j| \geq 2\ ,\ 1 \leq i, j \leq n-1\ .$$

$$(2)\ \tilde{\sigma}_i \cdot \tilde{\sigma}_{i+1} \cdot \tilde{\sigma}_i = \tilde{\sigma}_{i+1} \cdot \tilde{\sigma}_i \cdot \tilde{\sigma}_{i+1} \ \text{ for } \ 1 \leq i \leq n-2\ .$$

There is an obvious homomorphism

$$\iota_n : B_n \to B(n)\ ,$$

which maps $\tilde{\sigma}_i$ onto $\sigma_i$. The homomorphism $\iota_n$ is well defined, since the relations (1) and (2) hold in both $B_n$ and $B(n)$. We shall prove that $\iota_n$ is an isomorphism, and that will complete the proof of the theorem.

For $B(n)$ we have the braid group sequence

$$1 \to H(n) \xrightarrow{\rho_n} B(n) \xrightarrow{\tau_n} \Sigma_n \to 1\ ,$$

where $H(n)$ is the group of coloured braids (trivial permutation of strings), which we identify with $H(n) = \pi_1(F_n(\mathbb{E}^3), \bar{c}_0)$, and $\tau_n$ is the permutation homomorphism.

For the abstract group $B_n$, we have a natural surjective homomorphism

$$\tilde{\tau}_n : B_n \to \Sigma_n\ ,$$

defined by mapping $\tilde{\sigma}_i$ onto the transposition $(i, i+1) \in \Sigma_n$, i.e. the permutation that just interchanges $i$ and $i+1$. Since relations corresponding to (1) and (2) hold among the transpositions in $\Sigma_n$, the homomorphism $\tilde{\tau}_n$ is well defined, and since the transpositions generate $\Sigma_n$, it is indeed surjective.

Let $H_n = \ker \tilde{\tau}_n$ denote the kernel of $\tilde{\tau}_n$, and let $\tilde{\rho}_n\colon H_n \to B_n$ denote the inclusion homomorphism. For the abstract group $B_n$ we have then the short exact sequence

$$1 \to H_n \xrightarrow{\tilde{\rho}_n} B_n \xrightarrow{\tilde{\tau}_n} \Sigma_n \to 1 \ .$$

Since also $\tau_n(\sigma_i) = (i, i+1)$ is the transposition $(i, i+1) \in \Sigma_n$ , we get a commutative diagram of short exact sequences

$$\begin{array}{ccccccccc}
1 & \longrightarrow & H_n & \xrightarrow{\tilde{\rho}_n} & B_n & \xrightarrow{\tilde{\tau}_n} & \Sigma_n & \longrightarrow & 1 \\
 & & \downarrow{\scriptstyle \iota'_n} & & \downarrow{\scriptstyle \iota_n} & & \downarrow{\scriptstyle 1} & & \\
1 & \longrightarrow & H(n) & \xrightarrow[\rho_n]{} & B(n) & \xrightarrow[\tau_n]{} & \Sigma_n & \longrightarrow & 1 \ ,
\end{array}$$

where $\iota'_n$ is the restriction of $\iota_n$ to the subgroup $H_n$ in $B_n$ .

Using the 5–lemma we get then immediately

<u>Lemma 4.1.</u> The homomorphism $\iota_n : B_n \to B(n)$ is an isomorphism if and only if the homomorphism $\iota'_n : H_n \to H(n)$ is an isomorphism.

The idea is now to prove that $\iota'_n : H_n \to H(n)$ is an isomorphism, and then according to Lemma 4.1 we will have completed our task to prove that $B_n$ and $B(n)$ are isomorphic. The reduction from the full braid group to the subgroup of coloured braids has the advantage that one can apply induction over the number of strings since coloured braids have no permutation of strings.

To prove that $\iota'_n : H_n \to H(n)$ is an isomorphism, we need to know a presentation of $H_n$. Since this is a purely group theoretical problem, we shall at this place only state the final result, which is proved using the so–called Reidemeister – Schreier rewriting process. A proof is offered in Appendix 1 by Lars Gæde, pp. 153–170.

<u>Lemma 4.2.</u> The group $H_n$ admits a presentation with

$$\text{generators: } a_{ij} = \tilde{\sigma}_{j-1}\, \tilde{\sigma}_{j-2} \cdots \tilde{\sigma}_{i+1}\, \tilde{\sigma}_i^2\, \tilde{\sigma}_{i+1}^{-1} \cdots \tilde{\sigma}_{j-2}^{-1}\, \tilde{\sigma}_{j-1}^{-1} \quad \text{for } 1 \leq i < j \leq n$$

and defining relations:

$$a_{rs}^{-1}\, a_{ij}\, a_{rs} = \begin{cases} a_{ij} & \text{if } i < r < s < j \text{ or } r < s < i < j \\ a_{rj}\, a_{ij}\, a_{rj}^{-1} & \text{if } r < i = s < j \\ a_{rj}\, a_{sj}\, a_{ij}\, a_{sj}^{-1}\, a_{rj}^{-1} & \text{if } i = r < s < j \\ a_{rj}\, a_{sj}\, a_{rj}^{-1}\, a_{sj}^{-1}\, a_{ij}\, a_{sj}\, a_{rj}\, a_{sj}^{-1}\, a_{rj}^{-1} & \text{if } r < i < s < j \ . \end{cases}$$

[There appears to be a slight mistake – unimportant in most contexts – in the presentation of $H_n$ given in Birman's book, page 20.]

The geometric n – braid that corresponds to $a_{ij}$ , $1 \leq i < j \leq n$ , i.e. the braid $\sigma_{j-1}\,\sigma_{j-2} \cdots \sigma_{i+1}\,\sigma_i^2\,\sigma_{i+1}^{-1} \cdots \sigma_{j-2}^{-1}\,\sigma_{j-1}^{-1}$ , just takes the j string once around the i string and undercrosses intermediate strings twice in the standard projection shown in Figure 11.

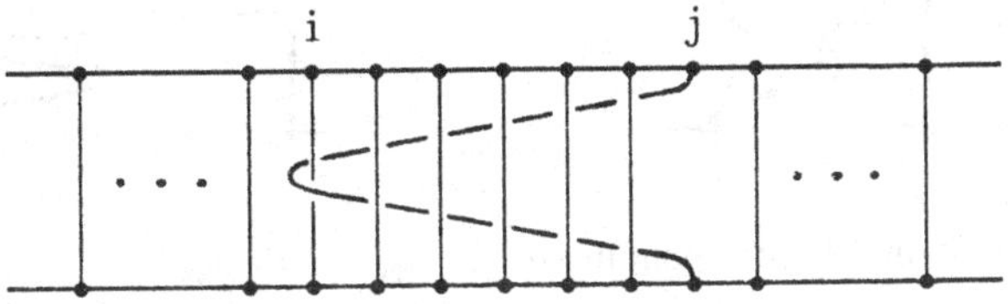


The geometric n–braid $\sigma_{j-1}\,\sigma_{j-2} \cdots \sigma_{i+1}\,\sigma_i^2\,\sigma_{i+1}^{-1} \cdots \sigma_{j-2}^{-1}\,\sigma_{j-1}^{-1}$

Figure 11

The inverse braid corresponding to $a_{ij}^{-1}$ does the same except for taking the j string the opposite way around the i string. It is then intuitively clear that a coloured braid can be resolved into a product of the braids corresponding to $a_{ij}$ and their inverses. [In particular, the reader should figure out: (1) how to take the i string around the j string for $i < j$ ; (2) how to take the j string once around the i string, overcrossing intermediate strings twice.] This process is called <u>combing the braid.</u> See Figure 12. In other words: The system of braids corresponding to the system $\{a_{ij} \mid 1 \leq i < j \leq n\}$ is a system of generators for $H(n)$.

The group $H_{n-1}$ can be regarded as the subgroup of $H_n$ that is generated by $\{a_{ij} \mid 1 \leq i < j \leq n-1\}$ . There is a natural homomorphism $\eta : H_n \to H_{n-1}$ defined by the rule $\eta(a_{ij}) = a_{ij}$ , if $1 \leq i < j \leq n-1$ , and $\eta(a_{in}) = 1$ , if $1 \leq i < n$ . The kernel of $\eta$ , ker $\eta$ , is then the normal closure in $H_n$ of the elements $a_{1n}, a_{2n}, \ldots, a_{(n-1)n}$ . The subgroup $U_n$ in $H_n$ generated by the elements $a_{1n}, a_{2n}, \ldots, a_{(n-1)n}$ is, however, already normal, since $a_{rs}\, a_{in}\, a_{rs}^{-1}$ for all relevant sets of indices belongs to $U_n$ according to the relations in Lemma 4.2. Therefore, ker $\eta = U_n$ . Altogether, we have the short exact sequence

$$1 \to U_n \to H_n \xrightarrow{\eta} H_{n-1} \to 1 \ .$$

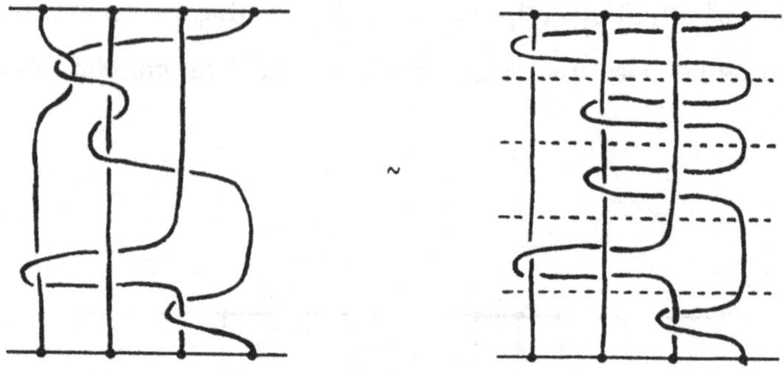

4–braid corresponding to $a_{14}\, a_{24}\, a_{24}\, a_{13}^{-1}\, a_{34}$

Figure 12

We shall now define a corresponding short exact sequence for the braid groups. For that purpose consider the fibration

$$\pi : F_n(\mathbb{E}^2) \to F_{n-1}(\mathbb{E}^2) \ ,$$

with fibre $F_{n-1,1}(\mathbb{E}^2) = \mathbb{E}^2 \backslash Q_{n-1}$ , defined in Theorem 2.1.

Since the homotopy groups of the spaces in the fibration $\pi : F_n(\mathbb{E}^2) \to F_{n-1}(\mathbb{E}^2)$ vanish in dimensions $\geq 2$ , the homotopy sequence for $\pi$ reduces to the short exact sequence

$$1 \to \pi_1(F_{n-1,1}(\mathbb{E}^2)) \to \pi_1(F_n(\mathbb{E}^2)) \xrightarrow{\pi_*} \pi_1(F_{n-1}(\mathbb{E}^2)) \to 1 \ .$$

For the coloured braid groups we have the identifications

$$H(n) = \pi_1(F_n(\mathbb{E}^2)) \ \text{ and } \ H(n-1) = \pi_1(F_{n-1}(\mathbb{E}^2)) \ ,$$

and thereby we get the short exact sequence

$$1 \to \pi_1(F_{n-1,1}(\mathbb{E}^2)) \to H(n) \xrightarrow{\pi_*} H(n-1) \to 1 \ .$$

It is obvious that $\pi_*$ on the level of braids operates by removing string number n from the n–braids. Also note that

$$\ker \pi_* = \pi_1(F_{n-1,1}(\mathbb{E}^2)) = \pi_1(\mathbb{E}^2 \backslash Q_{n-1})$$

is a free group on n–1 generators.

The geometric braid corresponding to $a_{ij}$ behaves under $\pi_*$ just as $a_{ij}$ behaves under $\eta$. Hence the following diagram is commutative,

$$\begin{array}{ccccccccc}
1 & \longrightarrow & U_n & \longrightarrow & H_n & \xrightarrow{\eta} & H_{n-1} & \longrightarrow & 1 \\
 & & \downarrow \iota''_n & & \downarrow \iota'_n & & \downarrow \iota'_{n-1} & & \\
1 & \longrightarrow & \pi_1(F_{n-1,1}(\mathbb{E}^2)) & \longrightarrow & H(n) & \xrightarrow[\pi_*]{} & H(n-1) & \longrightarrow & 1
\end{array}$$

where $\iota''_n$ is the restriction of $\iota'_n$ to $U_n$ .

Recall that $P_1 = (1,0)$, $P_2 = (2,0)$,...,$P_n = (n,0)$ is the set of points in $\mathbb{E}^2$ used as the set of initial points for the geometric n–braids. The coloured n–string braids considered as loops in $F_n(\mathbb{E}^2)$ are then based at the point $\bar{c}_0 = (P_1,...,P_n) \in F_n(\mathbb{E}^2)$. Correspondingly, the coloured (n–1)–string braids in H(n–1) are described by loops in $F_{n-1}(\mathbb{E}^2)$ based at the point $(P_1,...,P_{n-1}) \in F_{n-1}(\mathbb{E}^2)$ . The base point in the fibre $F_{n-1,1}(\mathbb{E}^2)$ for the fibration $\pi : F_n(\mathbb{E}^2) \to F_{n-1}(\mathbb{E}^2)$ , is then the point $P_n \in F_{n-1,1}(\mathbb{E}^2) = \mathbb{E}^2 \backslash \{P_1,...,P_{n-1}\}$ .

The analysis of the geometric braid $\iota'_n(a_{ij})$ corresponding to $a_{ij} \in H_n$ shows that it winds the j string once around the i string. It is therefore clear that $\iota''_n(a_{jn}) \in \pi_1(F_{n-1,1}(\mathbb{E}^2))$ , for $1 \leq j < n$ , is represented by the loop in

$F_{n-1,1}(\mathbb{E}^2) = \mathbb{E}^2\backslash\{P_1,\ldots,P_{n-1}\}$ based at $P_n$ that encircles the point $P_j$ once and separates it from the points $P_1,\ldots,P_{j-1},P_{j+1},\ldots,P_{n-1}$ . See Figure 13. It follows then immediately that the set of elements $\{\iota''_n(a_{jn}) \mid 1 \leq j < n\}$ is a free basis for the free group $\pi_1(F_{n-1,1}(\mathbb{E}^2)) = \pi_1(\mathbb{E}^2\backslash\{P_1,\ldots,P_{n-1}\})$ of rank $n-1$.

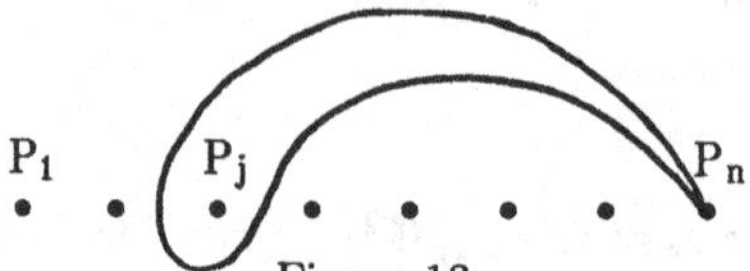

Figure 13

The group $U_n = \ker \eta$ is generated by the set of elements $\{a_{jn} \mid 1 \leq j < n\}$ , and as we have just seen, this set of generators is mapped onto a basis for the free group $\pi_1(F_{n-1,1}(\mathbb{E}^2))$ under $\iota''_n$ . Hence there can be no relations among these generators, since a free group of finite rank cannot be isomorphic to one of its proper factor groups (the so–called "Hopfian" property for finitely generated free groups [8], Theorem 2.13). In other words: the group $U_n$ is itself a free group of rank $n-1$, which is mapped isomorphically onto $\pi_1(F_{n-1,1}(\mathbb{E}^2))$ by $\iota''_n$ .

Now observe that $H_1 = 1$ and $\pi_1(F_{0,1}(\mathbb{E}^2)) = 1$ . Therefore $\iota'_1$ is an isomorphism. Assume inductively that $\iota'_{n-1}$ is an isomorphism. Then, since $\iota''_n$ is an isomorphism for all $n$ , the 5–lemma – applied to the above diagram of short exact sequences – implies that $\iota'_n$ is an isomorphism.

This completes the proof of Artin's presentation theorem for the braid group.

## 5. Representation of braids as automorphisms of free groups.

Our objective in this section is to prove that the braid group $B(n)$ is isomorphic to a subgroup of the automorphism group $Aut(\mathbb{F}_n)$ of the free group $\mathbb{F}_n$ on $n$ generators. This representation theorem for $B(n)$ was proved by Artin in his first paper on braid groups [22] in 1925. Before stating the theorem, we note that an automorphism of a free group is completely specified by its effect on a set of generators.

Theorem 5.1 (Artin Representation Theorem). Let $\mathbb{F}_n$ be the free group on a set of $n$ generators $x_1,...,x_n$ for a fixed integer $n \geq 1$, and let $Aut(\mathbb{F}_n)$ denote the group of (right) automorphisms on $\mathbb{F}_n$.

Then $B(n)$ is isomorphic to the subgroup of $Aut(\mathbb{F}_n)$ consisting of those right automorphisms $\overline{\beta}$ on $\mathbb{F}_n$ satisfying the following conditions

$$x_i\overline{\beta} = A_i\, x_{\tau(i)}\, A_i^{-1} \quad \text{for } 1 \leq i \leq n$$

and
$$(x_1...x_n)\overline{\beta} = x_1...x_n \;,$$

where $\tau$ is a permutation of $\{1,...,n\}$ and every $A_i$ belongs to $\mathbb{F}_n$, i.e. $A_i$ is a word in the generators $x_1,...,x_n$.

Under this isomorphism, the elementary braid $\sigma_i$, $1 \leq i \leq n-1$, corresponds to the automorphism $\overline{\sigma}_i \in Aut(\mathbb{F}_n)$ defined by

$$\begin{aligned} x_i\,\overline{\sigma}_i &= x_i\, x_{i+1}\, x_i^{-1} \\ x_{i+1}\,\overline{\sigma}_i &= x_i \\ x_j\,\overline{\sigma}_i &= x_j \quad \text{for all } j \neq i, i+1 \;, \end{aligned}$$

and the permutation $\tau$ for the automorphism $\overline{\beta} \in Aut(\mathbb{F}_n)$ corresponding to the braid $\beta \in B(n)$ is exactly the permutation of the braid.

Proof As usual we identify $\mathbb{E}^3$ with $\mathbb{R}^3$ and take the points

$$P_1 = (1,0,0)\;,\; P_2 = (2,0,0),...,P_n = (n,0,0) \text{ in the plane } z = 0$$

and
$$P_1' = (1,0,1)\;,\; P_2' = (2,0,1),...,P_n' = (n,0,1) \text{ in the plane } z = 1$$

as the set of end points for geometric n–braids.

Let $P_0 = (0,0,0)$ and $P'_0 = (0,0,1)$. Furthermore, let $\mathbb{E}^2_0$ denote the plane $z = 0$, and $\mathbb{E}^2_1$ the plane $z = 1$.

The free group $\mathbb{F}_n$ on n generators $x_1,\ldots,x_n$ can then be naturally identified with the fundamental groups of the punctured planes as follows:

$$\begin{aligned}\mathbb{F}_n &= \pi_1(\mathbb{E}^2_0 \setminus \{P_1,\ldots,P_n\}, P_0)\\ &= \pi_1(\mathbb{E}^2_1 \setminus \{P'_1,\ldots,P'_n\}, P'_0).\end{aligned}$$

The free generator $x_i$ is represented in $\pi_1(\mathbb{E}^2_0 \setminus \{P_1,\ldots,P_n\}, P_0)$ by the loop in $\mathbb{E}^2_0 \setminus \{P_1,\ldots,P_n\}$ at $P_0$ that just encircles $P_i$ once in the anticlockwise direction as shown in Figure 14.

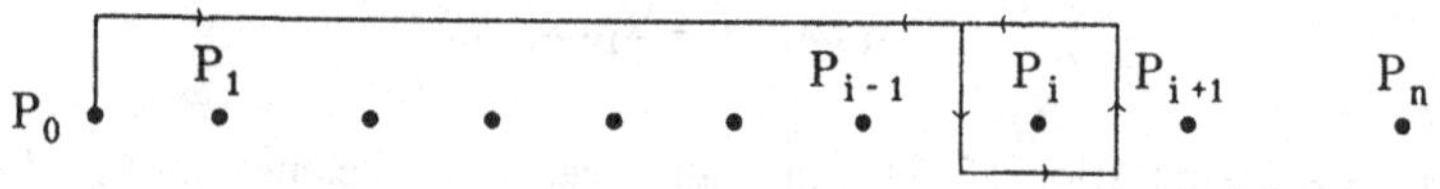


Figure 14

Similarly, $x_i$ has a representative in $\pi_1(\mathbb{E}^2_1 \setminus \{P'_1,\ldots,P'_n\}, P'_0)$.

Now, let $\beta$ be a geometric n–braid in $\mathbb{E}^3$. We shall consider $\mathbb{E}^3$ with the strings of the braid $\beta$ removed. Corresponding to $\beta$ we can then define a map

$$\bar{\beta} : \mathbb{F}_n = \pi_1(\mathbb{E}^2_0 \setminus \{P_1,\ldots,P_n\}, P_0) \to \mathbb{F}_n = \pi_1(\mathbb{E}^2_1 \setminus \{P'_1,\ldots,P'_n\}, P'_0),$$

by pushing a loop $\ell$ in $\mathbb{E}^2_0 \setminus \{P_1,\ldots,P_n\}$ at $P_0$ down along the gaps left by the n strings in $\beta$ to a loop $\ell\bar{\beta}$ in $\mathbb{E}^2_1 \setminus \{P'_1,\ldots,P'_n\}$ at $P'_0$.

This procedure really defines a single–valued mapping $\bar{\beta}$ on homotopy classes as indicated, since it is easy to prove

Assertion 1. If $\ell_1$ and $\ell_2$ are loops in $\mathbb{E}^2_0 \setminus \{P_1,\ldots,P_n\}$ at $P_0$, which are homotopic (relative to $P_0$), then $\ell_1\bar{\beta}$ and $\ell_2\bar{\beta}$ are homotopic in $\mathbb{E}^2_1 \setminus \{P'_1,\ldots,P'_n\}$ (relative to $P'_0$).

It is also easy to prove the following Assertion 2, which shows that $\bar{\beta}$ is a homomorphism.

<u>Assertion 2.</u> If $\ell_1$ and $\ell_2$ are loops in $\mathbb{E}_0^2 \setminus \{P_1,...,P_n\}$ at $P_0$, then the product loop $\ell_1 \cdot \ell_2$ satisfies

$$(\ell_1 \cdot \ell_2)\bar{\beta} = (\ell_1\bar{\beta}) \cdot (\ell_2\bar{\beta}) .$$

<u>Assertion 3.</u> $\bar{\beta}$ is a right automorphism.

To prove Assertion 3 we have to produce an inverse homomorphism to $\bar{\beta}$. But this is easy. Just use the pushing up procedure, where you push loops in $\mathbb{E}_1^2 \setminus \{P_1',...,P_n'\}$ at $P_0'$ up along the gaps left by the n strings in $\beta$ to loops in $\mathbb{E}_0^2 \setminus \{P_1,...,P_n\}$ at $P_0$.

From the definition of homotopy of braids we get

<u>Assertion 4.</u> If $\beta'$ is an n–braid homotopic to $\beta$, then $\bar{\beta}' = \bar{\beta}$ .

Assertion 4 ensures that we get a well defined map of $B(n)$ into $\mathrm{Aut}(\mathbb{F}_n)$ .

<u>Assertion 5.</u> The map $B(n) \to \mathrm{Aut}(\mathbb{F}_n)$, which to the braid $\beta \in B(n)$ associates the right automorphism $\bar{\beta} \in \mathrm{Aut}(\mathbb{F}_n)$ is a homomorphism. In other words: if $\beta_1$ and $\beta_2$ are geometric n–braids, then

$$\overline{\beta_1 \cdot \beta_2} = \bar{\beta}_1 \circ \bar{\beta}_2 .$$

Assertion 5 follows immediately from the definition of product of braids.

<u>Assertion 6.</u> The automorphism $\bar{\beta} \in \mathrm{Aut}(\mathbb{F}_n)$ corresponding to the n–braid $\beta \in B(n)$ satisfies

$$(x_1x_2...x_n)\bar{\beta} = x_1x_2...x_n .$$

The proof of Assertion 6 follows by observing that $x_1x_2...x_n$ is homotopic to a loop in $\mathbb{E}_0^2 \setminus \{P_1,...,P_n\}$ at $P_0$, which encircles the points $P_1, P_2,...,P_n$ once in the anti-clockwise direction. See Figure 15. Now $(x_1x_2...x_n)\bar{\beta}$ does the same in $\mathbb{E}_1^2 \setminus \{P_1',...,P_n'\}$ with respect to the points $P_1', P_2',...,P_n'$. Hence it is homotopic to $x_1x_2...x_n$ as a loop in $\mathbb{E}_1^2 \setminus \{P_1',...,P_n'\}$ at $P_0'$. This proves Assertion 6.

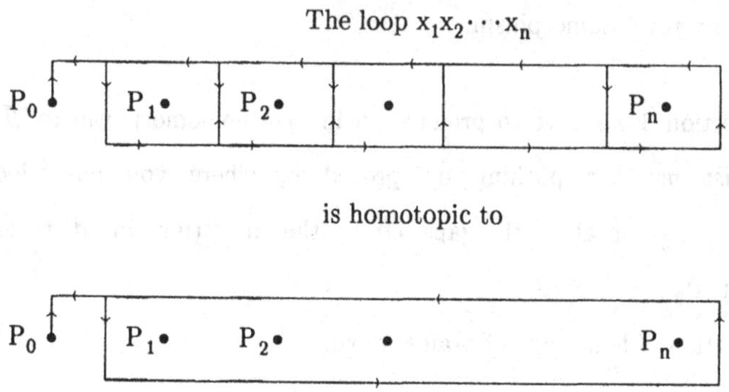


Figure 15

Evaluation of $\bar{\sigma}_i$ and $\overline{\sigma_i^{-1}}$

Let $\sigma_i$, $1 \le i \le n-1$, be the *i*th elementary geometric n–braid. Then clearly

$$x_j\bar{\sigma}_i = x_j \quad \text{for} \quad j \neq i, i+1 \ .$$

Since the $i+1$ string passes under the $i$ string in the standard projection of $\sigma_i$, it is also clear that

$$x_{i+1}\bar{\sigma}_i = x_i \ .$$

The $i$ string passes over the $i+1$ string in $\sigma_i$, and therefore, when we slide down $x_i$ along the braid $\sigma_i$, the loop defining $x_i\bar{\sigma}_i$ will pass in front of $P_i'$, as in Figure 16a. The loop in Figure 16a is homotopic to the loop in Figure 16b, which represents $x_i\, x_{i+1}\, x_i^{-1}$ . Consequently,

$$x_i\overline{\sigma}_i = x_i\, x_{i+1}\, x_i^{-1} \;.$$

This establish the formulas for $\overline{\sigma}_i$ stated in the theorem.

a)

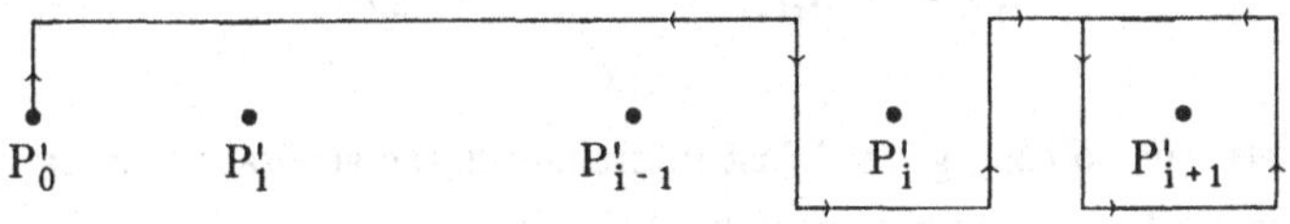


Loop representing $x_i\overline{\sigma}_i$

b)

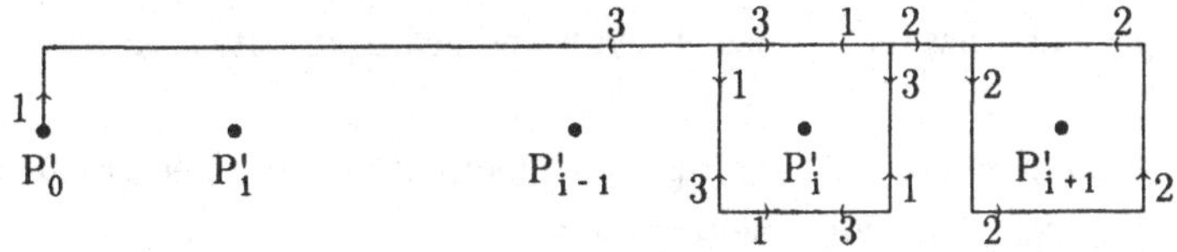


Loop representing $x_i\, x_{i+1}\, x_i^{-1}$

Arrows marked 1,2,3 correspond to the loops $x_i$ , $x_{i+1}$ , $x_i^{-1}$ , respectively, which you run through in that order.

Figure 16.

Since $\overline{\sigma_i^{-1}} = (\overline{\sigma}_i)^{-1}$ , a straightforward computation shows that

$$\begin{aligned} x_j\, \overline{\sigma_i^{-1}} &= x_j \quad \text{for} \quad j \neq i\,, i+1 \\ x_i\, \overline{\sigma_i^{-1}} &= x_{i+1} \\ x_{i+1}\, \overline{\sigma_i^{-1}} &= x_{i+1}^{-1}\, x_i\, x_{i+1} \;. \end{aligned}$$

<u>Evaluation of $\overline{\beta}$</u> (for an arbitrary braid $\beta \in B(n)$)

An arbitrary braid $\beta \in B(n)$ can be written as a product

$$\beta = \sigma_{i_1}^{\epsilon_1} \ldots \sigma_{i_k}^{\epsilon_k} \;\; \text{with} \;\; \epsilon_j = \pm 1 \;.$$

Then $\overline{\beta} = \overline{\sigma_{i_1}^{\epsilon_1}} \circ \ldots \circ \overline{\sigma_{i_k}^{\epsilon_k}}$ by Assertion 5.

From the above formulas for the action of $\overline{\sigma_i^{\pm 1}}$, it follows then immediately that there exist elements $A_i \in \mathbb{F}_n$, $1 \leq i \leq n$, and a permutation $\tau$ of $\{1,\ldots,n\}$ such that

$$x_i \overline{\beta} = A_i \, x_{\tau(i)} \, A_i^{-1} \quad \text{for} \quad 1 \leq i \leq n \; .$$

Since each $\overline{\sigma_i^{\pm 1}}$ (up to conjugation) just interchanges the generators $x_i$ and $x_{i+1}$ in $\mathbb{F}_n$, it is clear, that $\tau$ is the permutation of the braid $\beta$.

To complete the proof of Theorem 5.1 it now only remains to prove, that any automorphism in $\mathrm{Aut}(\mathbb{F}_n)$ satisfying the conditions in the theorem is realized by a unique braid in $B(n)$. This has independent interest and hence we state it as

<u>Lemma 5.2.</u> Suppose that $\xi$ is a right endomorphism of the free group $\mathbb{F}_n$ on n generators $x_1,\ldots,x_n$, satisfying the conditions

$$x_i \, \xi = A_i \, x_{\tau(i)} \, A_i^{-1} \quad \text{for} \quad 1 \leq i \leq n$$

and

$$(x_1 x_2 \ldots x_n) \, \xi = x_1 x_2 \ldots x_n$$

for elements $A_i \in \mathbb{F}_n$, $1 \leq i \leq n$, and a permutation $\tau$ of $\{1,\ldots,n\}$.

Then there exists a unique (equivalence class) geometric n–braid $\beta \in B(n)$ such that $\xi = \overline{\beta}$. In particular, $\xi$ must therefore be an automorphism of $\mathbb{F}_n$.

<u>Proof</u> The proof proceeds by induction over the integer $L(\xi) = \sum_{i=1}^{n} L(A_i)$, where $L(A_i)$ denotes the letter length of the word $A_i \in \mathbb{F}_n$.

If $L(\xi) = 0$, then every $A_i$ is the empty word, i.e. the identity element $1 \in \mathbb{F}_n$. Since $\xi$ leaves the product $x_1 x_2 \ldots x_n$ unaltered, the permutation $\tau$ must then also be the identity. Altogether, $\xi$ is therefore the identity automorphism $\mathrm{id}_{\mathbb{F}_n}$ on $\mathbb{F}_n$. From the definition of the automorphism $\overline{\beta}$ associated with a braid $\beta$, it is clear for topological reasons, that the only n–braid realizing $\mathrm{id}_{\mathbb{F}_n}$ must be the equivalence class of the trivial n–braid $\epsilon$. Hence if $L(\xi) = 0$ we must have $\xi = \overline{\epsilon}$.

Suppose now that the lemma is true for every endomorphism $\xi$ satisfying the above conditions, when $L(\xi) < m$ for $m \geq 1$, and let then $\xi$ be an endomorphism with $L(\xi) = m$.

The properties of $\xi$ imply that the following equality holds in $\mathbb{F}_n$

$$(*) \qquad (A_1 x_{\tau(1)} A_1^{-1}) \cdot (A_2 x_{\tau(2)} A_2^{-1}) \cdot \ldots \cdot (A_n x_{\tau(n)} A_n^{-1}) = x_1 x_2 \ldots x_n .$$

Since the right hand side of $(*)$ has letter length $n$, some cancellations must take place on the left hand side. (Otherwise $A_1 = \ldots = A_n = 1$ and this would be false since $L(\xi) > 0$.) We assume that the words $A_i x_{\tau(i)} A_i^{-1}$ are fully reduced.

<u>Assertion.</u> For such cancellations there are only two possibilities. There must exist some $i = 1,\ldots,n-1$, such that

either (a) $x_{\tau(i)} A_i^{-1}$ is absorbed by $A_{i+1}$, i.e.

$A_{i+1} = A_i x_{\tau(i)}^{-1} B_{i+1}$ for some $B_{i+1} \in \mathbb{F}_n$

or (b) $A_i^{-1}$ absorbs $A_{i+1} x_{\tau(i+1)}$, i.e.

$A_i = A_{i+1} x_{\tau(i+1)} C_i$ for some $C_i \in \mathbb{F}_n$ .

<u>Proof of assertion.</u> Suppose first that one of the terms $A_{i_0} x_{\tau(i_0)} A_{i_0}^{-1}$ for $1 \leq i_0 \leq n$, is completely absorbed by the other terms in the free cancellations which reduse LHS in $(*)$ to RHS. We ask how the letter $x_{\tau(i_0)}$ is absorbed by these cancellations? If $x_{\tau(i_0)}$ is absorbed by a letter to the left of $x_{\tau(i_0-1)}$, then (a) is satisfied for $i = i_0 - 1$. If $x_{\tau(i_0)}$ is absorbed by a letter in $A_{i_0-1}^{-1}$, then (b) is satisfied for $i = i_0 - 1$. If $x_{\tau(i_0)}$ is absorbed by a letter in $A_{i_0+1}$, then (a) is satisfied for $i = i_0$. If $x_{\tau(i_0)}$ is absorbed by a letter to the right of $x_{\tau(i_0+1)}$, then (b) is satisfied for $i = i_0$. Since $x_{\tau(i_0)}$ cannot be absorbed by either $x_{\tau(i_0-1)}$ or $x_{\tau(i_0+1)}$, since $\tau(i_0 - 1)$, $\tau(i_0)$ and $\tau(i_0 + 1)$ are different indexes, all possible cases of this first type have been treated.

Suppose next that no term $A_i x_{\tau(i)} A_i^{-1}$ is completely absorbed. In this case, some residue $R_i$ will remain for each $A_i x_{\tau(i)} A_i^{-1}$ after all free reductions have been made on LHS of (*). Then according to (*) $R_1R_2...R_n = x_1x_2...x_n$, which implies that $R_i = x_i$ for each $i = 1,...,n$. Now examine the term $A_1 x_{\tau(1)} A_1^{-1}$. The initial letter in this term can only be $x_1$. We consider first the case where $A_1 x_{\tau(1)} A_1^{-1}$ is not identic to $x_1$. Then $A_1 = x_1 \tilde{A}_1$ for a word $\tilde{A}_1 \in \mathbb{F}_n$, and

$$A_1 x_{\tau(1)} A_1^{-1} = x_1 \tilde{A}_1 x_{\tau(1)} A_1^{-1} .$$

Since $\tilde{A}_1 x_{\tau(1)} A_1^{-1}$ must be completely absorbed, there are two possibilities: Either $x_{\tau(1)}$ is absorbed by $A_2$, in which case (a) is satisfied for $i = 1$, or $x_{\tau(1)}$ is absorbed by a letter to the right of $x_{\tau(2)}$, in which case (b) is satisfied for $i = 1$. If on the other hand, $A_1 x_{\tau(1)} A_1^{-1} = x_1$, the entire argument can be repeated for $A_2 x_{\tau(2)} A_2^{-1}$, etc.

In this way we see that in every case either (a) or (b) is true. This proves the assertion.

We now continue with the proof of the lemma.

In case (a) we have that

$$L(\overline{\sigma}_i \xi) < L(\xi) ,$$

where $\overline{\sigma}_i$ is the automorphism of $\mathbb{F}_n$ induced by the elementary geometric n–braid $\sigma_i$, since

$$\begin{aligned}
x_j(\overline{\sigma}_i \xi) &= x_j \xi = A_j x_{\tau(j)} A_j^{-1} \quad \text{for } j \neq i, i+1 \\
x_i(\overline{\sigma}_i \xi) &= (x_i x_{i+1} x_i^{-1}) \xi = x_i \xi \cdot x_{i+1} \xi \cdot (x_i \xi)^{-1} \\
&= (A_i x_{\tau(i)} A_i^{-1}) \cdot (A_{i+1} x_{\tau(i+1)} A_{i+1}^{-1}) \cdot (A_i x_{\tau(i)}^{-1} A_i^{-1}) \\
&= (A_i B_{i+1}) x_{\tau(i+1)} (A_i B_{i+1})^{-1} \\
x_{i+1} (\overline{\sigma}_i \xi) &= x_i \xi = A_i x_{\tau(i)} A_i^{-1}
\end{aligned}$$

and $L(A_i B_{i+1}) = L(A_i) + L(B_{i+1}) = L(A_{i+1}) - 1$ .

In case (b) we have that

$$L(\overline{\sigma_i^{-1}}\,\xi) < L(\xi) \ ,$$

where $\overline{\sigma_i^{-1}}$ is the automorphism of $\mathbb{F}_n$ induced by the inverse of $\sigma_i$ , since

$$\begin{aligned}
x_j\,(\overline{\sigma_i^{-1}}\,\xi) &= x_j\,\xi = A_j\,x_{\tau(j)}\,A_j^{-1} \quad \text{for } j \neq i\,,\,i+1 \\
x_i\,(\overline{\sigma_i^{-1}}\,\xi) &= x_{i+1}\,\xi = A_{i+1}\,x_{\tau(i+1)}\,A_{i+1}^{-1} \\
x_{i+1}\,(\overline{\sigma_i^{-1}}\,\xi) &= (x_{i+1}^{-1}\,x_i\,x_{i+1})\,\xi = (x_{i+1}\,\xi)^{-1}\cdot(x_i\,\xi)\cdot(x_{i+1}\,\xi) \\
&= (A_{i+1}\,x_{\tau(i+1)}^{-1}\,A_{i+1}^{-1})\cdot(A_i\,x_{\tau(i)}\,A_i^{-1})\cdot(A_{i+1}\,x_{\tau(i+1)}\,A_{i+1}^{-1}) \\
&= (A_{i+1}\,C_i)\,x_{\tau(i)}\,(A_{i+1}\,C_i)^{-1}
\end{aligned}$$

and $L(A_{i+1} C_i) = L(A_{i+1}) + L(C_i) = L(A_i) - 1$ .

By the induction hypothesis, there exists therefore a unique (equivalence class) n–braid $\beta_0 \in B(n)$ such that in

$$\text{case (a):} \quad \overline{\sigma}_i\,\xi = \overline{\beta}_0 \qquad\qquad \text{case (b):} \quad \overline{\sigma_i^{-1}}\,\xi = \overline{\beta}_0 \ .$$

If we define $\beta \in B(n)$ by

$$\text{case (a):} \quad \beta = \sigma_i^{-1}\,\beta_0 \qquad\qquad \text{case (b):} \quad \beta = \sigma_i\,\beta_0 \ ,$$

then obviously $\overline{\beta} = \xi$ .

This proves Lemma 5.2 and, as already remarked, completes the proof of Theorem 5.1.

Remark 5.3. Theorem 5.1 provides immediately a solution to the word problem in $B(n)$ . For a group defined by generators and relations, the word problem is to find an algorithm to decide whether a given word represents the identity element. By representing the braids as automorphisms of the free group $\mathbb{F}_n$ , such an algorithm is easy to define.

Using Theorem 5.1, we can give a new geometric interpretation of the braid group B(n) .

Let $D^2$ denote the unit disc in $\mathbb{E}^2$ . The boundary of $D^2$ is the unit circle $S^1 = \partial D^2$ . Consider a set of n points $Q_n = \{q_1,...,q_n\}$ in the interior of $D^2$ . The points are arbitrarily chosen, but henceforward fixed. Choose a base point $q_0 \in S^1 = \partial D^2$ . Then the free group $\mathbb{F}_n$ on n generators $x_1,...,x_n$ can be identified with the fundamental group of $D^2 \setminus Q_n$ , i.e.

$$\mathbb{F}_n = \pi_1(D^2 \setminus Q_n , q_0) = \pi_1(D^2 \setminus \{q_1,...,q_n\} , q_0) .$$

We get a representative for $x_i$ by taking a loop in $D^2$ based at $q_0$ , which encircles the point $q_i$ once anticlockwise and avoids the points $q_j$ for $j \neq i$ . See Figure 17.

Let $\mathcal{B}_n(D^2, S^1)$ denote the space of homeomorphisms $h: D^2 \to D^2$ of $D^2$ , which keep $S^1 = \partial D^2$ pointwise fixed and permute the points in $Q_n$ among themselves. A homeomorphism $h: D^2 \to D^2$ in $\mathcal{B}_n(D^2, S^1)$ satisfies in other words the conditions $h(p) = p$ for $p \in S^1$ and $h(q_i) = q_{\tau(i)}$ for $i = 1,...,n$ , where $\tau$ is some permutation of $\{1,...,n\}$ . We may consider $\mathcal{B}_n(D^2, S^1)$ as a topological space by giving it the compact – open topology.

Observe that each homeomorphism $h: D^2 \to D^2$ in $\mathcal{B}_n(D^2, S^1)$ preserves orientation.

Each homeomorphism $h \in \mathcal{B}_n(D^2, S^1)$ induces an automorphism $h_*$ of $\pi_1(D^2 \setminus Q_n , q_0)$ onto itself, or with the above identification, an automorphism $h_*: \mathbb{F}_n \to \mathbb{F}_n$ .

Theorem 5.4. The braid group B(n) can be identified with the group of automorphisms of $\mathbb{F}_n = \pi_1(D^2 \setminus Q_n)$ induced by the homeomorphisms $h \in \mathcal{B}_n(D^2, S^1)$ .

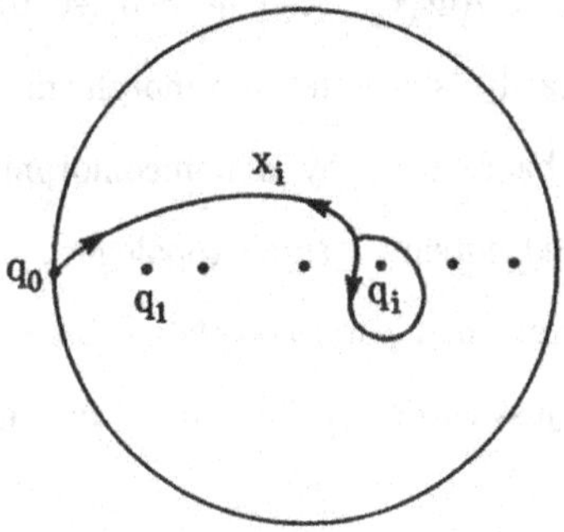


Figure 17

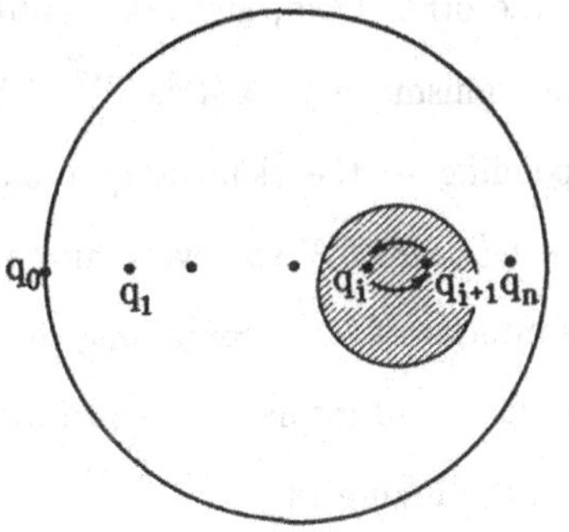


Figure 18

<u>Proof</u> A homeomorphism $h: D^2 \to D^2$ in $\mathscr{B}_n(D^2, S^1)$ will map a loop, which encircles $q_i$ once anticlockwise, onto a loop, which encircles the point $q_{\tau(i)} = h(q_i)$ once anticlockwise, since $h$ is orientation preserving. The image under $h$ of the path from $q_0$ to the loop around $q_i$ in the representative for $x_i$, combines with the path in the representative for $x_{\tau(i)}$ to define a loop in $D^2 \setminus Q_n$. The homotopy class of that loop defines an element $A_i \in \mathbb{F}_n$ such that

$$h_*(x_i) = A_i\, x_{\tau(i)}\, A_i^{-1} .$$

As in the proof of Theorem 5.1, it can be proved that the element $x_1...x_n$ is represented by a loop in $D^2 \setminus Q_n$ which encircles all the points $q_1,...,q_n$ once in the anticlockwise direction. Since this loop under $h$ is mapped onto a loop with the same property we get

$$h_*(x_1...x_n) = x_1...x_n .$$

Hence, by Theorem 5.1, there is a unique (equivalence class) braid $\beta \in B(n)$, such that $\bar{\beta} = h_* \in \mathrm{Aut}(\mathbb{F}_n)$.

On the other hand, any braid automorphism $\bar{\beta} \in \mathrm{Aut}(\mathbb{F}_n)$ can be realized by a homeomorphism $h \in \mathscr{B}_n(D^2, S^1)$. This follows easily, since the automorphism $\bar{\sigma}_i$ corresponding to the elementary braid $\sigma_i$ can be realized by a homeomorphism $h: D^2 \to D^2$, which keeps everything pointwise fixed outside a small topological disc (in the interior of $D^2$) containing the points $q_i$ and $q_{i+1}$, and no other point from the set $Q_n$, and inside the small topological disc just turns $q_i$ into $q_{i+1}$ and $q_{i+1}$ into $q_i$. See Figure 18.

This proves Theorem 5.4.

Remark 5.5. Let $h: D^2 \to D^2$ be a homeomorphism in $\mathscr{B}_n(D^2, S^1)$. Then we can get a geometric braid $\beta \in B(n)$ from $h$ as follows. By a classical theorem of Alexander from 1923 – often referred to as the Alexander trick – any homeomorphism $h: D^2 \to D^2$, which keeps the boundary $S^1 = \partial D^2$ pointwise fixed, is isotopic to the identity map through such homeomorphisms. See Lemma 5.6 below. To the given homeomorphism $h \in \mathscr{B}_n(D^2, S^1)$, we can therefore construct a map $H: D^2 \times [0,1] \to D^2 \times [0,1]$ with the following properties: (i) For each $t \in [0,1]$, $H$ maps the level $D^2 \times \{t\}$ homeomorphically onto itself; (ii) The subspaces $D^2 \times \{0\}$ and $S^1 \times [0,1]$ are kept pointwise fixed; (iii) When restricted to $D^2 \times \{1\}$, $H$ defines the given homeomorphism $h \in \mathscr{B}_n(D^2, S^1)$. The image of $\{q_1,\dots,q_n\} \times [0,1]$ under $H$ is then the geometric braid $\beta \in B(n)$ asked for.

For the convenience of the reader we insert a proof of the theorem of Alexander [20] just mentioned.

Lemma 5.6. (Alexander trick). Let $D^n$ denote the unit ball in euclidean n–space $\mathbb{E}^n$ and $S^{n-1} = \partial D^n$ its boundary (n–1)–sphere. Then any homeomorphism $h: D^n \to D^n$, which fixes $S^{n-1}$ pointwise, is isotopic to the identity map under an isotopy, which fixes $S^{n-1}$ pointwise. If $h(0) = 0$, then the isotopy may be chosen to fix $0 \in D^n$.

<u>Proof</u> Consider the cylinder $D^n \times [0,1]$ in euclidean $(n+1)$-space $\mathbb{E}^{n+1}$ with coordinates $(x,t) \in D^n \times [0,1]$. Let $C$ be the cone in $D^n \times [0,1]$ with top point $(0,0) \in D^n \times [0,1]$ and base $D^n \times \{1\}$. See Figure 19.

Let $h\colon D^n \to D^n$ be a homeomorphism, which fixes $S^{n-1} = \partial D^n$ pointwise. We consider $h$ as a homeomorphism on $D^n \times \{1\}$. Define $H\colon D^n \times [0,1] \to D^n \times [0,1]$ as follows.

Outside the cone $C$ and on $D^n \times \{0\}$ we take $H$ to be the identity map. Inside $C$, we define $H$ by mapping the ball $(D^n \times \{t\}) \cap C$ of radius $t$ for each $0 < t \le 1$ homeomorphically onto itself: first projecting it linearly onto $D^n \times \{1\}$ from the top of the cone, then mapping $D^n \times \{1\}$ onto itself by $h$ and finally projecting back to $(D^n \times \{t\}) \cap C$. Clearly, $H$ so defined is continuous, and when restricted to each level $D^n \times \{t\}$, $0 \le t \le 1$, it defines a homeomorphism of $D^n \times \{t\}$ onto itself.

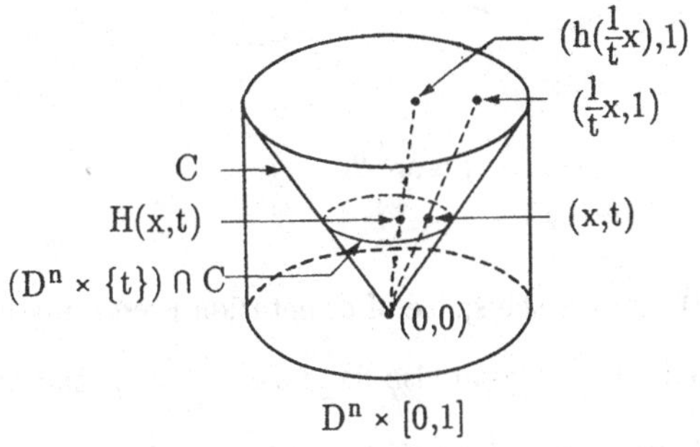


Figure 19

In coordinates we have

$$H(x,t) = \begin{cases} (x,t) & \text{for } t \le \|x\| \\ (t\, h(\tfrac{1}{t}x),\, t) & \text{for } t > \|x\| . \end{cases}$$

The map $H\colon D^n \times [0,1] \to D^n \times [0,1]$ is the isotopy desired. This proves Lemma 5.6.

Pursuing Remark 5.5 further, one can identify the group of (path–) components in $\mathcal{B}_n(D^2, S^1)$ with the braid group $B(n)$. This shows that $B(n)$ is a particular case of a mapping class group.

Consider in general an oriented surface $T_g$ of genus $g \geq 0$. Remove the interior of $r \geq 0$ embedded discs from $T_g$, and mark $n \geq 0$ points in the interior of the remaining space. Denote the resulting space by $T_g^{r,n}$. See Figure 20.

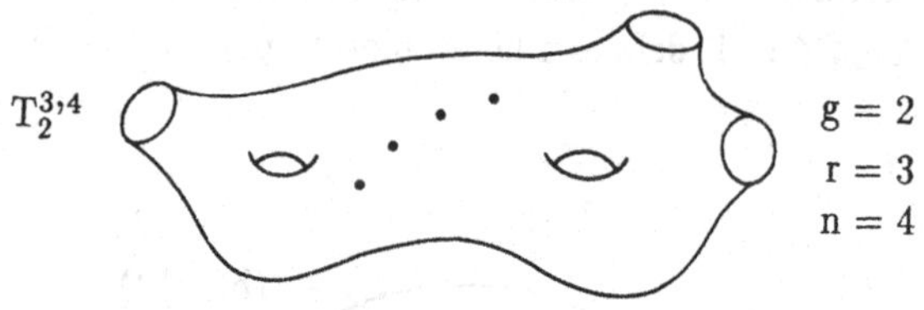


Figure 20

Let $\mathcal{B}_n(T_g^{r,n}, \partial T_g^{r,n})$ denote the space of orientation preserving homeomorphisms $h\colon T_g^{r,n} \to T_g^{r,n}$ that keep $\partial T_g^{r,n}$ pointwise fixed and permute the $n$ marked points on $T_g^{r,n}$ among themselves. Equip this space with the compact – open topology. The set of (path–) components in $\mathcal{B}_n(T_g^{r,n}, \partial T_g^{r,n})$ has a natural group structure induced by composition of homeomorphisms. This group $M(g, r, n) = \pi_0(\mathcal{B}_n(T_g^{r,n}, \partial T_g^{r,n}))$ is the general <u>mapping class group</u>. Note that $M(0, 1, n) = B(n)$. For further information about the mapping class group see Birman [1].

6. The Dirac string problem.

This section is slightly off the main path of our investigations of the Artin braid groups in the euclidean plane, but it does provide an interesting application of the theory of braids in the 2–sphere.

Take a solid object in 3–space (e.g. a wrench, a bottle opener, or as in Figure 21 a pair of scissors) and attach it to two posts (e.g. two table legs) by loose, or elastic strings, say with one string from one end of the object and two strings from the other end of the object to the two posts respectively.

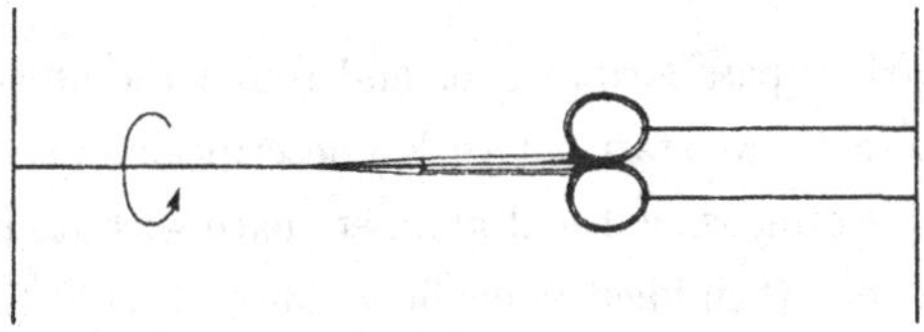

Figure 21

If you turn the object $2\pi$ around the "string–axis", the two strings going out to one of the posts will be twisted together. If you keep the object fixed, it is hardly a surprise, that it is impossible to remove the twisting of the strings by passing them over and round the object. However, if you turn the object another $2\pi$ around the "string–axis", so that you have now altogether made two full turns, the twisting becomes more complicated – but only apparently – for surprisingly enough, it is now possible to remove the twisting of the strings by passing them over and round the object.

The above phenomenon, or rather its explanation, is known as the Dirac string problem.

The experiment with a pair of scissors was performed by the Nobel laureate in physics P.A.M. Dirac at lectures in the 1930's to demonstrate the notion half–spin of elementary particles. Elementary particles with half–spin are called Fermions, and they include the electron, the proton and the neutron among others. During his work in the 1920's to establish relativistic quantum mechanical models for the elementary particles, Dirac had predicted, that if you turn a Fermion one full turn around an axis then it is not the same particle any more – it is in a different state. But if you turn it two full turns around an axis, then it is back to normal. For the neutron this was only verified experimentally as late as 1975. That a Fermion in this way is connected to the surrounding space in a nontrivial topological way seems at first to contradict our intuition about objects in space. Dirac's experiment shows that similar phenomena take place also in the macroscopical world.

The Dirac string problem was solved by M.H.A. Newman [40] in 1942, except that he used both an algebraic and a geometric description of braids in the 2–sphere $S^2$, and these descriptions were not formally known to be equivalent at that time. The equivalence of the two descriptions of braids in $S^2$ was first established by E. Fadell and J. Van Buskirk in the paper [29] from 1962.

We shall present an alternative purely geometric solution to the Dirac string problem due to E. Fadell [28].

Since we are allowed to pass strings over and round the object, we need to work with braids in $S^2$. Therefore we start out with a description of such braids.

Let $\mathbb{R}^3$ denote the 3–dimensional real number space with its usual euclidean inner product and norm. We may then identify euclidean 3–space $\mathbb{E}^3$ with $\mathbb{R}^3$.

Let $S_0 = S^2$ be the standard unit sphere in $\mathbb{R}^3$, and let $S_1$ be the sphere of radius 2. The shell in $\mathbb{R}^3$ between $S_0$ and $S_1$ can be identified with the product $S^2 \times [0,1]$. Let $S_t$, $0 \leq t \leq 1$, be the intermediate sphere with radius $1 + t$ between $S_0$ and $S_1$.

Choose n pairwise different points $P_1, P_2, \ldots, P_n$ on the inner sphere $S_0$, and project them radially onto the points $P'_1, P'_2, \ldots, P'_n$ on the outer sphere $S_1$. The case $n = 4$ is shown in Figure 22. In analogy with the definitions of Artin braids (geometric braids) in §1 and Fox braids in §3, we have the following definitions.

A <u>braid</u> $\beta$ <u>on</u> n <u>strings in</u> $S^2$ is

by <u>Artin's definition</u>: A system of n embedded arcs

$\mathcal{A} = \{\mathcal{A}_1, \ldots, \mathcal{A}_n\}$ in $S^2 \times [0,1]$ connecting the set of points $P_1, \ldots, P_n$ on $S_0$ to the set of points $P'_1, \ldots, P'_n$ on $S_1$, such that

(i) Each arc $\mathcal{A}_i$ intersects each intermediate sphere $S_t$ exactly once.

(ii) The arcs $\mathcal{A}_1, \ldots, \mathcal{A}_n$ intersect each intermediate sphere $S_t$ in exactly n different points.

by Fox's definition: A loop $f: [0,1] \rightarrow C_n(S^2)$ with $f(0) = f(1) = [P_1,...,P_n]$ , where $[P_1,...,P_n]$ is the unordered configuration of the points $P_1,...,P_n$ .

As usual we have a permutation of the braid, and we have a suitably defined notion of homotopy of braids. As before we do not distinguish between homotopic braids.

Exactly as in § 3, it can be shown that the two definitions coincide. Also the set of braids on n strings in $S^2$ has a natural group structure. A braid on 4 strings in $S^2$ is pictured in Figure 22.

A coloured (or pure) braid on n strings in $S^2$ is defined as the homotopy class of a loop $f: [0,1] \rightarrow F_n(S^2)$ with $f(0) = f(1) = (P_1,...,P_n)$ .

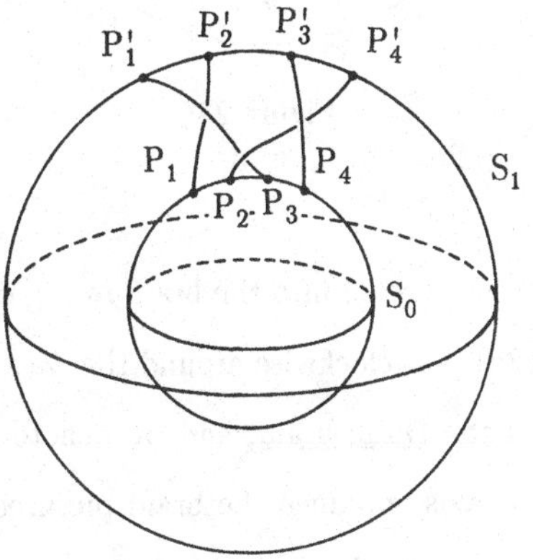


Figure 22

To obtain a model for the Dirac string problem, we use $S_0$ as our object. The sphere $S_0$ is then free to rotate about the origin $0 \in \mathbb{R}^3$ , while the sphere $S_1$ is kept fixed. We attach $S_0$ to $S_1$ via elastic strings as in Figure 23.

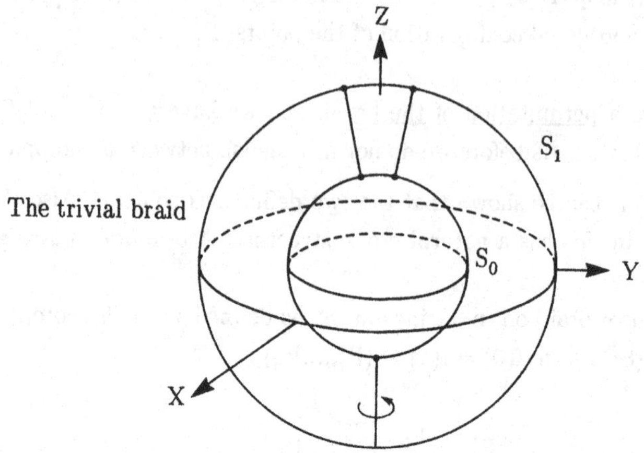


Figure 23

Turn the trivial braid in Figure 23 into the braid in Figure 24a by giving the inner sphere $S_0$ one full turn $(2\pi)$ anticlockwise around the Z–axis. The pure braid in $S^2$ obtained this way is called the Dirac braid, and we denote it by $\Delta$ . Two full turns $(4\pi)$ of $S_0$ around the Z–axis produce the braid pictured in Figure 24b, which is easily seen to be the product braid $\Delta^2$ in $S^2$ .

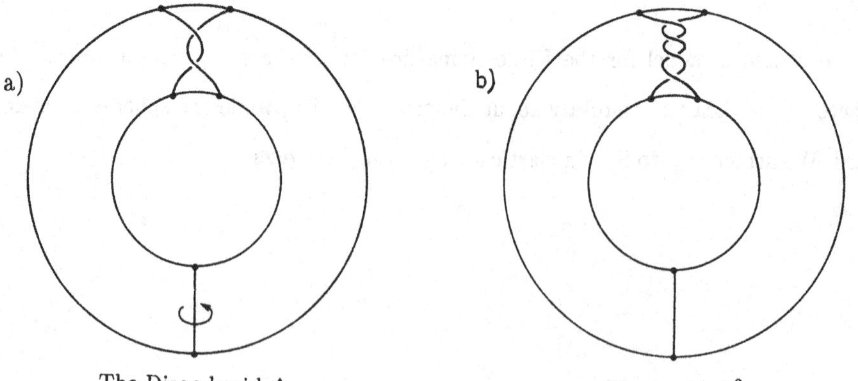


The Dirac braid $\Delta$ The braid $\Delta^2$

Figure 24

The solution to the Dirac string problem is now simply as follows: The Dirac braid $\Delta$ is not homotopic to the trivial braid in $S^2$, whereas $\Delta^2$ is. This is contained in the following

Theorem 6.1. Let $\Delta$ be the Dirac braid in $S^2$ and let $m \geq 1$ be an integer. Then $\Delta^m$ is a nontrivial braid in $S^2$ for $m$ odd and the trivial braid in $S^2$ for $m$ even.

Before entering the proof we will need some preparations.

Let $V_{3,2}$ denote the Stiefel manifold of orthonormal 2–frames in $\mathbb{R}^3$. A point in $V_{3,2}$ is then a pair $(v_1, v_2)$ of orthogonal unit vectors in $\mathbb{R}^3$.

It is well known that the map $\pi'\colon V_{3,2} \to S^2$, which projects $(v_1, v_2) \in V_{3,2}$ onto $v_1 \in S^2$, is a locally trivial fibration. The fibre of $\pi'$ over $v_1 = (0,0,1)$ can obviously be identified with the unit circle

$$S^1 = \{v_2 \in \mathbb{R}^3 \mid \|v_2\| = 1 \text{ and } v_1 \perp v_2\}.$$

From Theorem 2.1, we have the locally trivial fibration $\pi\colon F_3(S^2) \to S^2$, which projects $(x_1, x_2, x_3) \in F_3(S^2)$ onto $x_1 \in S^2$. The fibre of $\pi$ over $v_1 = (0,0,1)$ is then the configuration space $F_2(S^2 \setminus (0,0,1))$.

By stereographical projection from $(0,0,1)$ of $S^2 \setminus (0,0,1)$ onto $\mathbb{R}^2 \times \{0\} \subset \mathbb{R}^3$, the space $F_2(S^2 \setminus (0,0,1))$ can be identified with $F_2(\mathbb{R}^2)$. On the other hand, we have a natural homeomorphism of $F_2(\mathbb{R}^2)$ onto $\mathbb{R}^2 \times (\mathbb{R}^2 \setminus \{0\})$ defined by mapping $(x_2, x_3)$ onto $(x_2 + x_3, x_2 - x_3)$. This homeomorphism induces a homotopy equivalence from $F_2(\mathbb{R}^2)$ to $S^1$, which maps

$$(x_2, x_3) \in F_2(\mathbb{R}^2) \text{ onto } \frac{x_2 - x_3}{\|x_2 - x_3\|} \in S^1.$$

Altogether, we conclude that the fibre of $\pi\colon F_3(S^2) \to S^2$ over $v_1 = (0,0,1)$ is homotopy equivalent to $S^1$.

Lemma 6.2. For any fixed angle $\alpha \in ]0,\pi[$ , the map $\varphi_\alpha : V_{3,2} \to F_3(S^2)$ , defined by

$$\varphi_\alpha(v_1, v_2) = (v_1, \cos\alpha\, v_1 + \sin\alpha\, v_2 , \cos\alpha\, v_1 - \sin\alpha\, v_2) ,$$

is a fibre homotopy equivalence,

$$\begin{array}{ccc} V_{3,2} & \xrightarrow{\varphi_\alpha} & F_3(S^2) \\ & {\scriptstyle\pi'}\searrow \quad \swarrow{\scriptstyle\pi} & \\ & S^2 & \end{array} .$$

Proof. Obviously, $\varphi_\alpha$ commutes with the projections $\pi'$ and $\pi$ onto $S^2$ , and hence it is a fibre map. Consider the restriction of $\varphi_\alpha$ to the fibres of $\pi'$ and $\pi$ over $v_1 = (0,0,1)$ . With the above identifications of fibres, it is easy to see that $\varphi_\alpha$ maps $v_2 \in S^1 = (\pi')^{-1}(v_1)$ identically onto $v_2 \in S^1 \simeq \pi^{-1}(v_1)$ . Therefore, $\varphi_\alpha$ is a homotopy equivalence on fibres. By a well known theorem of Dold ([25], Theorem 6.3), $\varphi_\alpha$ is then a fibre homotopy equivalence. This proves the lemma.

Let $SO(3)$ denote the group of orientation preserving, linear isometries in $\mathbb{R}^3$ . Then there is a canonical homeomorphism $h: SO(3) \to V_{3,2}$ , which maps the isometry $A \in SO(3)$ onto the 2-frame $(v_1, v_2) = (A(0,0,-1) , A(0,1,0))$.

On the other hand, $SO(3)$ can be identified with the set of positions of $S_0$ , since an isometry turns $S^2$ into a position of $S_0$ , and conversely, corresponding to each position of $S_0$ , there is exactly one isometry, which turns $S^2$ into that position of $S_0$ .

It is well known ([13], § 22.3) that the fundamental group of $SO(3)$ is the cyclic group of order 2,

$$\pi_1(SO(3)) \simeq \mathbb{Z}/2 ,$$

and that a generator of $\pi_1(SO(3))$ is defined by the loop

$$\zeta: [0,1] \to SO(3) , \zeta(0) = \zeta(1) = I ,$$

based at the identity isometry I , which corresponds to one full turn of $S^2 = S_0$ around the Z–axis.

With these preparations we are ready for the

Proof of Theorem 6.1. It is easy to check that under the homotopy equivalence (for a suitable angle $\alpha \in ]\frac{\pi}{2}, \pi[$)

$$SO(3) \xrightarrow[\cong]{h} V_{3,2} \xrightarrow[\simeq]{\varphi_\alpha} F_3(S^2) ,$$

the loop $\zeta$: $[0,1] \to SO(3)$ generating $\pi_1(SO(3)) \cong \mathbb{Z}/2$ is mapped onto the loop f: $[0,1] \to F_3(S^2)$ describing the Dirac braid $\Delta$ . Since $\zeta^m$ is nontrivial for m odd and trivial for m even , the theorem follows.

The following series of pictures shows how to untwist the strings for the braid $\Delta^2$ in Figure 24b.

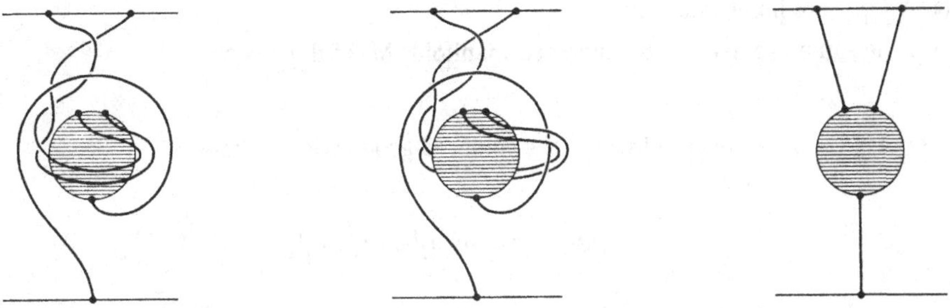

Figure 25

To untwist $\Delta^m$ for $m \geq 4$ even , we just have to repeat $\frac{m}{2}$ times the procedure in Figure 25.

Also observe that the number of strings do not matter as long as we attach our object $S_0$ to $S_1$ by more than 3 strings. This is so, since if we cannot untwist 3 strings, more strings would just cause more difficulties, and the untwisting procedure described in Figure 25 works independently of the number of strings.

## EXERCISES

1. Prove that every geometric n–braid is homotopic to an n–braid in which the strings are polygonal arcs and the crossings of arcs in the standard plane projection of the braid are transversal. Also prove that it may be assumed (up to equivalence of braids) that crossings of strings occur on different levels.

2. Complete the missing details in the proof that the set of geometric n–braids $B(n)$ form a group.

3. Let $D^2 = \{y \in \mathbb{E}^2 \mid \|y\| \leq 1\}$ be the unit disc in the euclidean plane $\mathbb{E}^2$. Let $q$ and $q'$ be points in the interior of $D^2$. Prove that there exists a homeomorphism $h: D^2 \to D^2$, which fixes the boundary circle $S^1 = \partial D^2$ pointwise, and maps $q$ into $q'$.

Let $Q_m = \{q_1,\dots,q_m\}$ and $Q'_m = \{q'_1,\dots,q'_m\}$ be two subsets in $\mathbb{E}^2$, each containing $m$ points. Prove that there exists a homeomorphism $h: \mathbb{E}^2 \to \mathbb{E}^2$ such that $h(q_i) = q'_i$ for each $i = 1,\dots,m$. Use this to prove that the configuration spaces $F_n(\mathbb{E}^2 \setminus Q_m)$ and $F_n(\mathbb{E}^2 \setminus Q'_m)$ are homeomorphic.

Do the same problem for a connected manifold $M$ of dimension $\geq 2$.

4. For $1 \leq i < j \leq n$ denote by $\bar{a}_{ij}$ the coloured geometric n–braid

$$\bar{a}_{ij} = \sigma_{j-1}\, \sigma_{j-2} \cdots \sigma_{i+1}\, \sigma_i^2\, \sigma_{i+1}^{-1} \cdots \sigma_{j-2}^{-1}\, \sigma_{j-1}^{-1} .$$

(i) Write the following braid on 4 strings as a product of the braids $\bar{a}_{12}, \bar{a}_{13}, \bar{a}_{14}, \bar{a}_{23}, \bar{a}_{24}, \bar{a}_{34}$ and their inverses:

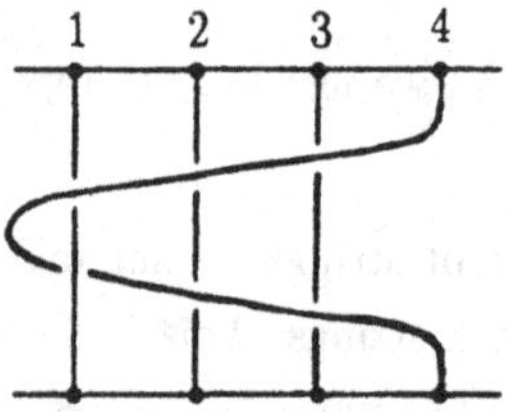

(ii) Prove geometrically that the system of coloured braids $\{\bar{a}_{ij} \mid 1 \leq i < j \leq n\}$ generates $H(n)$ .

5. In the unit disc $D^2$ in $\mathbb{E}^2$ consider the points $q_1 = (-\frac{1}{2}, 0)$ and $q_2 = (\frac{1}{2}, 0)$ . Construct an explicit homeomorphism $h: D^2 \to D^2$ , which fixes the boundary circle $S^1 = \partial D^2$ pointwise and maps $q_1$ into $q_2$ and $q_2$ into $q_1$ , and which represents the elementary 2–braid $\sigma_1$ .

6. Prove that $B(3)$ admits the presentation $\langle a,b \mid a^3 = b^2 \rangle$ , e.g. by setting $a = \sigma_1 \sigma_2$ and $b = \sigma_1 \sigma_2 \sigma_1$ .

7. Prove that there is a unique surjective homomorphism $\varphi: B(n) \to \mathbb{Z}$ onto the integers $\mathbb{Z}$ , such that

$$\varphi(\sigma_{i_1}^{\nu_1} \ldots \sigma_{i_k}^{\nu_k}) = \sum_{j=1}^{k} \nu_j$$

for every braid word $\sigma_{i_1}^{\nu_1} \ldots \sigma_{i_k}^{\nu_k}$ in $B(n)$ .

Prove that the kernel of $\varphi$ is the commutator subgroup of $B(n)$ .

8. Let $\sigma = \sigma_1 \sigma_2 \ldots \sigma_{n-1}$ be the product of the elementary n–braids. Prove that $B(n)$ is generated by $\sigma_1$ and $\sigma$ .

9. Prove that the coloured braid group $H(n)$ is the smallest normal subgroup of $B(n)$ that contains all the squares $\sigma_1^2, \ldots, \sigma_{n-1}^2$ of the elementary n–braids.

10. Let $S^r$ denote the r–sphere in euclidean $(r+1)$–space $\mathbb{E}^{r+1}$ , $r \geq 2$ , and $V_{r+1,2}$ the Stiefel manifold of orthonormal 2–frames in $\mathbb{E}^{r+1}$ . There are locally trivial fibrations

$$\pi: F_n(S^r) \to S^r \text{ , defined by } \pi(x_1, \ldots, x_n) = x_1$$

$$\pi': V_{r+1,2} \to S^r \text{ , defined by } \pi(v_1, v_2) = v_1 \text{ .}$$

(i) Prove that $\pi\colon F_2(S^r) \to S^r$ is a homotopy equivalence.

(ii) ([28], Theorem 2.4). Construct a fibre homotopy equivalence,

$$\begin{array}{ccc} V_{r+1,2} & \xrightarrow{\varphi} & F_3(S^r) \\ & {}_{\pi'}\searrow \quad \swarrow_{\pi} & \\ & S^r & . \end{array}$$

11. Let $M$ be a connected manifold of dimension $r \geq 2$, and let $Q_m = \{q_1,\ldots,q_m\}$ be a fixed set of $m$ distinct points in $M$. Set $Q_k = \{q_1,\ldots,q_k\}$ for $1 \leq k < m$, and let $Q_0 = \emptyset$ be the empty set.

(i) ([30], Theorem 1). Prove that the locally trivial fibration

$$\pi\colon F_{m,n}(M) \to M\backslash Q_m \text{ , defined by } \pi(x_1,\ldots,x_n) = x_1 \text{ ,}$$

admits a section if $m \geq 1$.

(ii) ([30], Theorem 2). Prove that

$$\pi_i(F_{1,n-1}(M)) = \sum_{k=1}^{n-1} \pi_i(M\backslash Q_k) \quad \text{(direct sum)}$$

for $i \geq 2$. Prove that if also $\pi\colon F_{0,n}(M) \to M$ admits a section, then

$$\pi_i(F_{0,n}(M)) = \sum_{k=0}^{n-1} \pi_i(M\backslash Q_k) \quad , \ i \geq 2 \text{ .}$$

(iii) ([30], Corollary 2.1). Prove that if $M = \mathbb{E}^r$ is euclidean r–space, then

$$\pi_i(F_{0,n}(\mathbb{E}^r)) = \sum_{k=1}^{n-1} \pi_i(\underbrace{S^{r-1}\vee\ldots\vee S^{r-1}}_{k}) \quad , \ i \geq 2 \text{ ,}$$

where $S^{r-1}$ is the $(r-1)$–sphere and $\vee$ denotes wedge product (one point union) of spaces.

Chapter II

# BRAIDS AND LINKS

The original motivation for Artin to introduce braids was their potential applications in the study of links. In this chapter we present therefore a few such applications. First we prove that every link can be obtained by closing a braid. This result is implicitely contained in a paper of Alexander from 1923. Then we give an introduction to a highly nontrivial and difficult theorem announced by Markov 1935, but apparently first proved in complete detail by Birman in her book from 1974. Markov's theorem turns the topological problem of classifying link types in euclidean 3–space into an algebraic problem involving the family of Artin braid groups for all numbers of strings. Next we present a proof due to Birman of a presentation theorem stated by Artin 1925 for the group of a link. Finally, we show how to obtain braid representations for links and give an example of computing the corresponding group of a link.

In the last few years there has been a tremendous development in the theory of links not the least due to the discovery of a new polynomial invariant for links made by V.F.R. Jones in 1985. Jones used a representation of the braid group to the group of units in certain Hecke algebras to define his (Laurent) polynomial. The Jones polynomial has by now been generalized to a two variable (Laurent) polynomial and a completely elementary approach to such polynomials has been found. A short guide to the literature and a very pleasing introduction to this subject is given in the following paper by two of the principal investigators:

> W.B.R. Lickorish and K.C. Millett: The New Polynomial Invariants of Knots and Links. Mathematics Magazine 61(1988), 3–23.

A new approach to the theorem of Alexander, and in particular the theorem of Markov, has been developed by Hugh Morton using his beautiful idea of treading knot diagrams. I am indepted to Hugh Morton for allowing me to reprint his paper "Treading knot diagrams", Math. Proc. Camb. Philos. Soc. 99(1986), 247–260, in Appendix 2.

1. Constructing links from braids.

Let $\beta$ be a geometric braid on n strings in euclidian 3–space $\mathbb{E}^3$. In this chapter we will for the sake of convenience draw the plane projections of a braid horizontally as in Figure 1, which pictures the 3–braid $\beta = \sigma_2 \sigma_1^2$ .

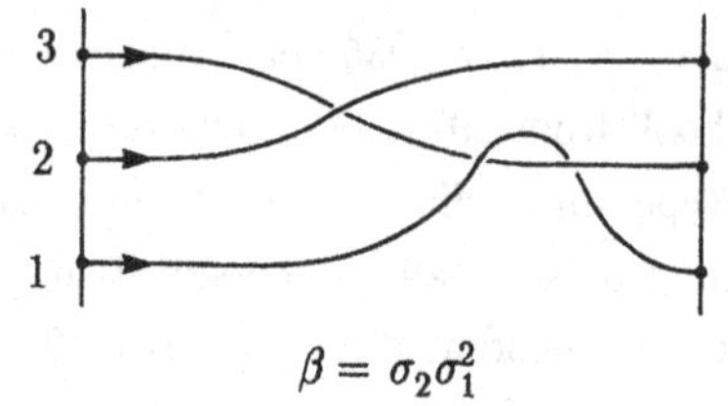


Figure 1

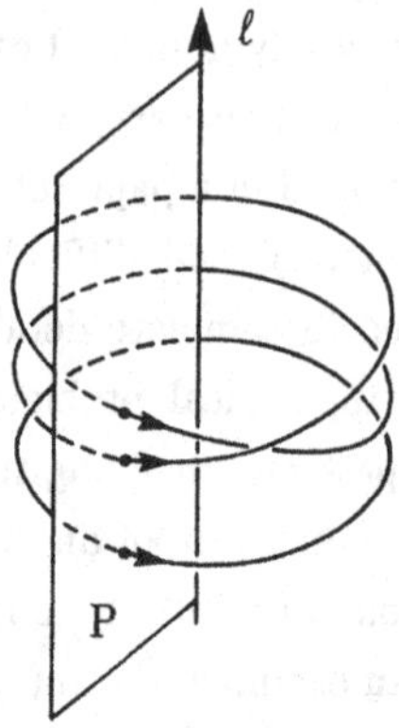


Figure 2

Now let $\ell$ be a suitable vertical axis placed behind the braid in $\mathbb{E}^3$ and close the braid around the axis by identifying the initial points and the end points of the braid. Corresponding to the braid in Figure 1 we get the closed braid in Figure 2.

We can give the strings in the braid $\beta$ an orientation such that in the projection of the braid, as in Figure 1, we move along the projections of the strings from left to right. This orientation of strings induces an orientation of the closed braid, and we can then give the axis $\ell$ an orientation, such that, when a point moves along the closed braid in the positive direction, it winds around the axis on a right hand screw. See Figure 2.

If conversely we take the closed braid in Figure 2 and cut it open along a half–plane P determined by $\ell$, and then fold it out, we get back the braid $\beta$ in Figure 1.

When we close the braid $\beta = \sigma_2 \sigma_1^2$ on 3 strings pictured in Figure 1, we get two interlocking embedded circles in $\mathbb{E}^3$ as pictured in Figure 2. Such a system of embedded circles in $\mathbb{E}^3$ is called a link – in our case a link with 2 components, or a 2–link. The general definition is as follows.

Definition 1.1. A link with m components (or an m–link) V in $\mathbb{E}^3$ is the union of $m \geq 1$ mutually disjoint, simple, closed polygonal curves, embedded in $\mathbb{E}^3$. The case $m = 1$ is called a knot.

As indicated, the mutually disjoint, simple, closed, polygonal curves, which make up a link, are called the components of the link. We take the components to be polygonal curves in order to avoid topological pathologies.

If a link is obtained by closing a braid $\beta$ as in Figure 2, then the number of components in the link must be equal to the number of cycles in the permutation of the braid.

The line segments on a link are called edges and their end points vertices. We shall use the notation [a b] for an edge on a link with vertices a and b.

Let V be a link in $\mathbb{E}^3$, and let [a b] be an edge on V with vertices a and b. Let c be a point in $\mathbb{E}^3$ different from a and b, and denote the (possibly degenerate) triangle spanned by a, b and c by $\Delta(a, b, c)$. Suppose that $\Delta(a, b, c)$ intersects the link V in exactly the edge [a b], i.e.

$$V \cap \Delta(a, b, c) = [a\ b] .$$

Then we say that the elementary deformation $\mathscr{E} = \mathscr{E}^{c}_{ab}$ is applicable to V, and we define a new link $\mathscr{E}^{c}_{ab}V$ by

$$\mathscr{E}^{c}_{ab}V = (V \setminus [a\ b]) \cup [a\ c] \cup [c\ b] .$$

In other words: To obtain $\mathscr{E}^{c}_{ab}V$ we deform V along the triangle $\Delta(a, b, c)$ by removing the edge [a b] and substituting with the two new edges $[a\ c] \cup [c\ b]$. See Figure 3.

If conversely, $[a\ c] \cup [c\ b]$ are consequtive edges on a link V' such that $V' \cap \Delta(a, b, c) = [a\ c] \cup [c\ b]$, then we can perform the inverse elementary deformation $\mathscr{E}^{-1} = (\mathscr{E}^{c}_{ab})^{-1}$ on V', substituting the edges $[a\ c] \cup [c\ b]$ by [a b].

As already indicated, we shall refer to both $\mathscr{E}^{c}_{ab}$ and its inverse $(\mathscr{E}^{c}_{ab})^{-1}$ as an elementary deformation.

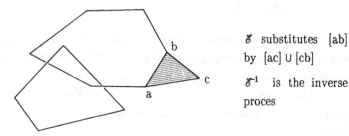

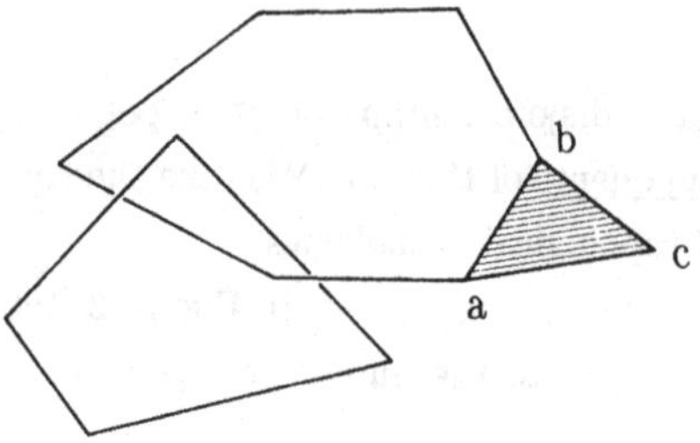


$\mathscr{E}$ substitutes [ab] by [ac] ∪ [cb]

$\mathscr{E}^{-1}$ is the inverse proces

Figure 3

We can now define an equivalence relation among links.

Definition 1.2. Two links V and V' in $\mathbb{E}^3$ are said to be combinatorially equivalent, if one can be transformed into the other by a finite sequence of elementary deformations.

It is easy to prove, that if two links V and V' are combinatorially equivalent, then they are ambient isotopic in $\mathbb{E}^3$ , i.e. there exists a continuous family of homeomorphisms $h_t\colon \mathbb{E}^3 \to \mathbb{E}^3$ on $\mathbb{E}^3$ , $0 \leq t \leq 1$ , such that $h_0 = id_{\mathbb{E}^3}$ is the identity map and such that $h_1(V) = V'$ . In other words: We can continuously deform V in 3–space (through links) to obtain V' . Conversely, it is also true that if V and V' are ambient isotopic links in $\mathbb{E}^3$ , then they are combinatorially equivalent. Equivalence of links in $\mathbb{E}^3$ corresponds therefore to the intuitive notions of being able to deform V continuously into V' through links.

It is obvious, that equivalent links have the same number of components.

The link in Figure 2, defined by closing the braid on 3 strings $\beta = \sigma_2\,\sigma_1^2$, is equivalent to the link defined by closing the braid $\beta' = \sigma_1^2$ on 2 strings pictured in Figure 4.

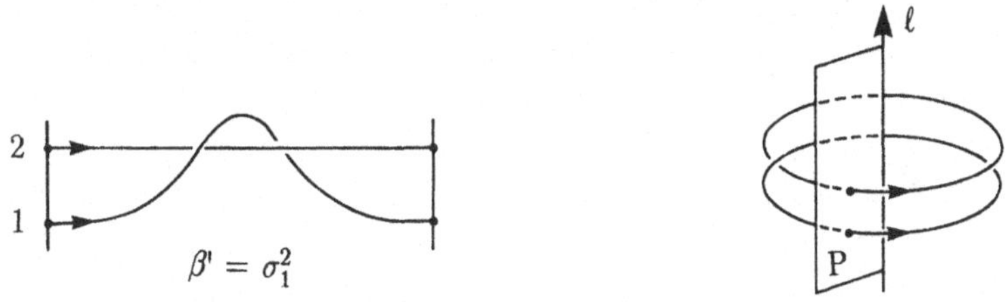


Figure 4

To prove the equivalence of the links in Figure 2 and Figure 4, just "untwist" the upper "circle" in Figure 2.

From this example we learn that two braids on a different number of strings may lead to equivalent links. There are strong restrictions on how this can happen. In fact, according to a theorem of Markov – to which we shall return in §3 – basically this can happen only, if one braid arises from the other by adding or deleting a string which shows up as a free factor in the factorizations of the braids in elementary braids, just as $\sigma_2$ shows up in $\beta = \sigma_2\,\sigma_1^2$ compared to $\beta' = \sigma_1^2$.

Another operation on braids, which leads to equivalent links, is conjugation. To be specific, if $\beta \in B(n)$ and $\gamma \in B(n)$, then the links we obtain by closing $\beta$ and $\gamma\,\beta\,\gamma^{-1}$ are equivalent. This is obvious, since when closing $\gamma\,\beta\,\gamma^{-1}$, the braid $\gamma$ will be attached to its inverse $\gamma^{-1}$, and hence this piece of the link can in a trivial way be unravelled whereby we get the link defined by $\beta$. See Figure 5.

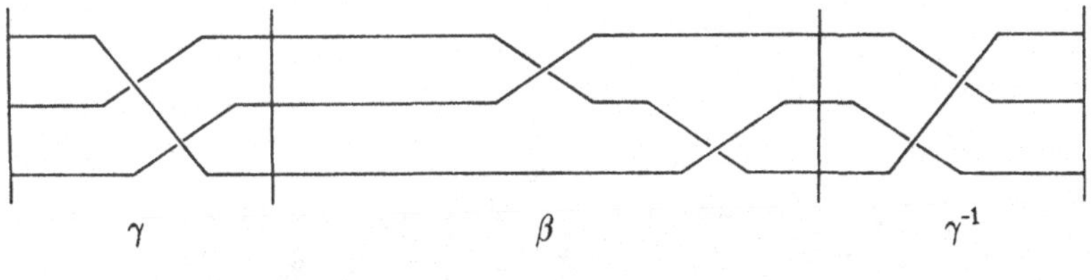


Figure 5

The theorem of Markov states roughly, that in order for two braids to define equivalent links, it is necessary and sufficient that one braid arises from the other by a finite sequence of moves, either involving adding or deleting a string which shows up as a free factor $\sigma_i^{\pm 1}$, or by a conjugation.

## 2. Representing link types by closed braids. A theorem of Alexander.

In this section we shall prove that every link is combinatorially equivalent to a link obtained by closing a braid. This result is due to Alexander, who in the paper [19] from 1923 proved a theorem on systems of knotted curves, which basically amounts to the above statement. We need some preparations.

Consider a link V in $\mathbb{E}^3$. Let $\ell$ be an arbitrary, but henceforth fixed, line in $\mathbb{E}^3$, which does not meet the link V. We shall refer to $\ell$ as the axis for V.

Definition 2.1. The link V is said to be in general position with respect to the axis $\ell$ if none of its edges are coplaner with $\ell$.

As the following lemma shows, we only have to worry about links in general position.

Lemma 2.2. Every link is combinatorially equivalent to some link in general position.

Proof. If [a b] is an edge on V, which is coplaner with the axis $\ell$, then we can choose a point c in $\mathbb{E}^3$ outside this plane and change [a b] to [a c] ∪ [c b] by an elementary deformation. If c is chosen sufficiently close to the edge [a b], then obviously $\mathscr{E}_{ab}^{c}$ will be an applicable elementary deformation. This procedure can be carried out for all the (finitely many) edges on V, which are coplaner with $\ell$, without interfering with the rest of the link. This proves the lemma.

We suppose from now on that the link V in $\mathbb{E}^3$ is in general position with respect to the axis $\ell$.

Suppose furthermore, that V is oriented, i.e. all the components of the link are oriented, simple, closed polygonal curves in $\mathbb{E}^3$. In other words: Each component of the link has a preferred direction attached. An orientation of the link assigns an orientation to every edge of the link. Note, that every link admits an orientation.

Fix also an orientation of the axis $\ell$.

The orientations of the link V and the axis $\ell$ enable us to divide the edges on V in positive and negative edges. We call an edge [a b] on V positive, when the half-plane P determined by $\ell$ and a point on [a b] turns on a right hand screw around $\ell$, when the point on [a b] moves along the edge in the positive direction determined by the orientation of V. See Figure 6. We write $[a\ b] > 0$. Call [a b] negative, when the half–plane P turns on a left hand screw around $\ell$, when [a b] is traversed in the positive direction. In this case, we write [a b] < 0 .

We define the <u>height</u> of V , denoted h(V) , as the number of negative edges on V.

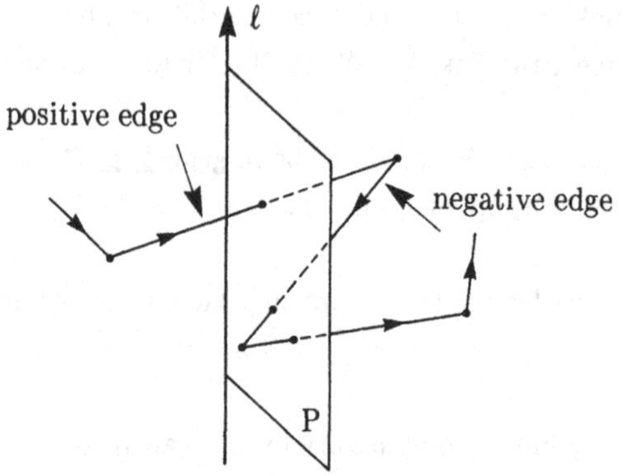


Figure 6

We can now characterize the links arising by closing a braid as in § 1.

<u>Lemma 2.3.</u> A link V in $\mathbb{E}^3$ can be constructed by closing a braid if and only if it admits an orientation and an oriented axis $\ell$, such that all edges on V are positive.

<u>Proof.</u> If V arises by closing a braid, we have already observed in § 1, that it admits an orientation and an orientation of the axis $\ell$ as required. Hence the conditions are necessary.

Suppose on the other hand that V is an m–link, which admits an orientation and an oriented axis $\ell$, such that all edges on V are positive. Cut V along a half–plane P determined by $\ell$. Since all edges on V are positive, the half–plane P must cut V in a finite number of points n (in general $n \neq m$), which remains constant, if the half–plane P is turned around the axis $\ell$. By following the paths of these n points in the half–plane P , when it is turned around the axis $\ell$, a braid $\beta$ is traced out. Closing this braid $\beta$ around $\ell$ gives us back V . Hence the conditions are also sufficient.

Motivated by Lemma 2.3 we make the following

Definition 2.4. A closed braid in $\mathbb{E}^3$ is an oriented link V in $\mathbb{E}^3$, which admits an oriented axis $\ell$ with respect to which all edges are positive.

By Lemma 2.3, the closed braids are then exactly the links in $\mathbb{E}^3$, which arise by closing (open) braids as in § 1. Hence the name is well chosen. The closed braid obtained from the open braid $\beta \in B(n)$ will be denoted by $\hat{\beta}$.

Now we can state and prove the theorem of Alexander, which is implicit in the paper [19] from 1923.

Theorem 2.5. Every link V in $\mathbb{E}^3$ is combinatorially equivalent to a closed braid.

Proof. Choose an orientation of the link V and suppose – as we may, according to Lemma 2.2 – that it is in general position with respect to an oriented axis $\ell$. If all edges of V are positive we are done. Otherwise, we shall remove negative edges one by one as follows.

Suppose that [a b] is a negative edge. First consider the special case, where we can find a point c in $\mathbb{E}^3$ such that the elementary deformation $\mathscr{E}_{ab}^{c}$ is applicable to V, i.e. $\Delta(a, b, c) \cap V = [a\ b]$, and such that the axis $\ell$ intersects the triangle $\Delta(a, b, c)$ in an interior point of the triangle. As is clear from Figure 7, the link

$$V' = \mathscr{E}_{ab}^{c} V = (V \setminus [a\ b]) \cup [a\ c] \cup [c\ b]$$

has then one less negative edge than V.

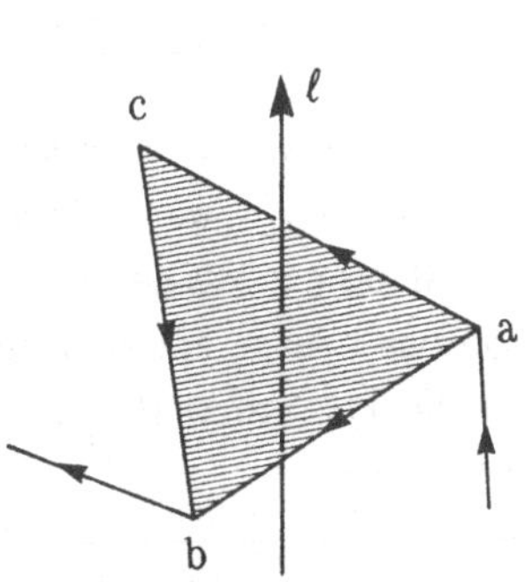


Figure 7

If we cannot find a single point c as above, we errect a <u>sawtooth</u> on [a b] as follows. Chop the edge [a b] up in small subedges $[a_{i-1}\ a_i]$ , $i = 1,\dots,k$ , with $a_0 = a$ and $a_k = b$ . Such a subdivision of the edge [a b] can obviously be done by applicable elementary deformations and hence does not change the combinatorial equivalence class of the link. Call the new link V' . If the subedges $[a_{i-1}\ a_i]$ are sufficiently small, then we can choose a point $c_i$ in $\mathbb{E}^3$ for each $i = 1,\dots,k$ , such that $\Delta(a_{i-1}, a_i, c_i) \cap V' = [a_{i-1}\ a_i]$ and such that $\ell$ intersects $\Delta(a_{i-1}, a_i, c_i)$ in an interior point. (To prove this, just observe that from each point of [a b] we can draw a line segment which intersects $\ell$ and avoids V except for the initial point. Such a line segment can be opened up to a small triangle with base on [a b] and with the desired properties.) Thereby we get a sawtooth as in Figure 8.

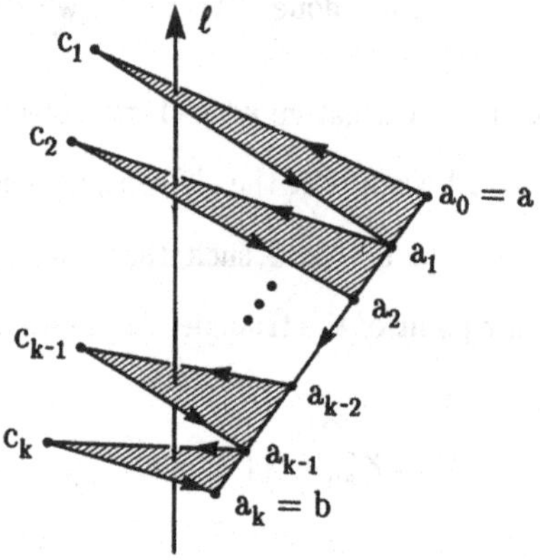


**Figure 8**

By the procedure in the special case we can now substitute the negative edges $[a_{i-1}\ a_i]$ on V' by positive edges, and thereby we arrive at a new link V" , which is combinatorially equivalent to V and has one less negative edge than V .

Now we remove negative edges one by one, and in a finite number of steps we arrive at a link with positive edges only. This proves the theorem.

## 3. Combinatorial equivalence of closed braids. Markov's theorem.

In this section we shall present the theorem of A.A. Markov on combinatorial equivalence of closed braids. The theorem was announced by Markov [38] in 1935, but a detailed proof was never published. The first complete proof is given by Joan Birman in her book [1] published 1974. It is a highly nontrivial theorem, and the proof is rather delicate. We shall not attempt to give all the details of the proof, but limit ourselves to describe the contents of the theorem and to give an indication why it is true. For an alternative proof see the paper by Hugh Morton reprinted in Appendix 2.

Let V be an (oriented) closed braid with (oriented) axis $\ell$. As in Definition 2.4, all the edges on V are then positive with respect to $\ell$.

First we shall describe two simple types of deformations through positive edges, which may be applied to V.

### A deformation of type $\mathscr{R}$.

Suppose that $[a\ b] > 0$ is a positive edge on V, and let c be a point in $\mathbb{E}^3$ such that $\Delta(a, b, c) \cap V = [a\ b]$ and such that $[a\ c] > 0$ and $[c\ b] > 0$ with respect to $\ell$. Then the elementary deformation $\mathscr{E}^c_{ab}$ is applicable to V. In $V' = \mathscr{E}^c_{ab}V$, we have removed the positive edge $[a\ b]$ from V and substituted it with the two new positive edges $[a\ c]$ and $[c\ b]$. See Figure 9. The elementary deformation $\mathscr{E}^c_{ab}$ and its inverse deformation is called a deformation of type $\mathscr{R}$.

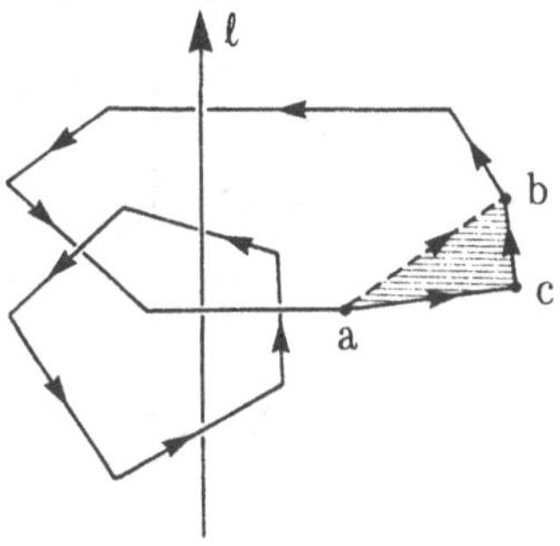


A deformation of type $\mathscr{R}$ removes one positive edge from V and substitutes it with two new positive edges; or vice versa.

Figure 9

A deformation of type $\mathscr{W}$.

Suppose that $[a\ d] > 0$ is a positive edge on $V$, and let $b$ be a point in $\mathbb{E}^3$ such that $\Delta(a, d, b) \cap V = [ad]$ and such that one of the edges $[a\ b]$ or $[b\ d]$ is positive and the other one negative with respect to $\ell$. Then the elementary deformation $\mathscr{E}_{ad}^{b}$ is applicable to $V$. In $V' = \mathscr{E}_{ad}^{b}V$, we have removed the positive edge $[a\ d]$ from $V$ and substituted it with the two new edges $[a\ b]$ and $[b\ d]$, one of which is positive and the other one negative with respect to $\ell$. Say, that $[b\ d]$ is negative. Let $c$ be a point in $\mathbb{E}^3$ such that $\Delta(b, d, c) \cap V' = [b\ d]$ and such that $[b\ c] > 0$ and $[c\ d] > 0$ with respect to $\ell$. Then the elementary deformation $\mathscr{E}_{bd}^{c}$ is applicable to $V'$. In the combination of two deformations, $\mathscr{E}_{ad}^{b}$ followed by $\mathscr{E}_{bd}^{c}$, we arrive at the closed braid

$$V'' = \mathscr{E}_{bd}^{c}\mathscr{E}_{ad}^{b}V = (V \backslash [a\ d]) \cup [a\ b] \cup [b\ c] \cup [c\ d],$$

in which we have substituted the positive edge $[a\ d]$ in $V$ with 3 new positive edges $[a\ b]$, $[b\ c]$ and $[c\ d]$. See Figure 10a. In Figure 10b we have pictured the situation when $[a\ b] < 0$ and $[b\ d] > 0$. Such a combined deformation, or its inverse, is called a deformation of type $\mathscr{W}$.

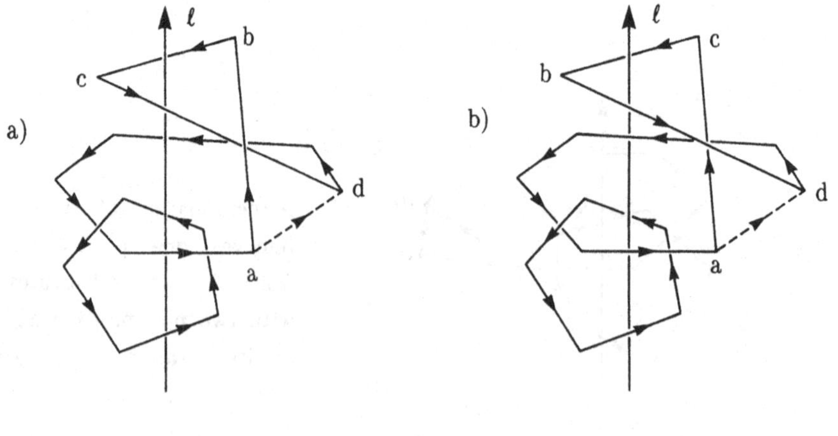


A deformation of type $\mathscr{W}$ removes one positive edge from $V$ and substitutes it with 3 new positive edges; or vice versa.

Figure 10

Geometrically we can think of a deformation of type $\mathscr{W}$ as pulling [a d] out along a rectangle, giving the rectangle a wrench, and then placing the resulting loop around the axis; or vice versa.

With these two types of deformations at our disposal, we may now formulate the geometric version of the theorem announced by A.A. Markov in 1935.

Theorem 3.1. (Geometric version). Let V and V* be two closed braids in $\mathbb{E}^3$, which are combinatorially equivalent. Then there exists a finite sequence of closed braids in $\mathbb{E}^3$, $V = V_0, V_1, V_2, \ldots, V_{s-1}, V_s = V^*$, such that for each $0 \leq i < s$, $V_{i+1}$ is obtained from $V_i$ by a single application of a deformation of type $\mathscr{R}$ or of type $\mathscr{W}$. (The converse is trivially true.)

Before giving a sketch of the proof of Theorem 3.1, we will make some comments.

By the theorem of Alexander (Theorem 2.5) we know that every link in $\mathbb{E}^3$ is combinatorially equivalent to a closed braid. Hence Theorem 3.1 is in fact a statement about combinatorial equivalence of two links in $\mathbb{E}^3$ in complete generality. We just have to choose orientations of the two links and then perform certain preliminary elementary deformations in order to deform them into oriented links with positive edges, and – if we like – even with respect to the same positive axis.

On the other hand we know by Lemma 2.3, that a closed braid V in $\mathbb{E}^3$ arises by closing an open braid $\beta \in B(n)$ around an axis. We write $V = \hat{\beta}$. The braid $\beta$ is however highly non–unique. In fact, as we have seen in § 1, not even the number of strings (the string index) in a representing braid $\beta$ is unique. Therefore we shall now specify the number of strings in a braid $\beta \in B(n)$ by writing $(\beta, n)$. By analysing the algebraic analogous of the deformations of type $\mathscr{R}$ or $\mathscr{W}$, the theorem of Markov can be given the following algebraic formulation.

Theorem 3.1. (Algebraic version). Let $\hat{\beta}$ and $\hat{\beta}^*$ be two closed braids in $\mathbb{E}^3$, with braid representatives $(\beta, n)$ and $(\beta^*, n^*)$. Then $\hat{\beta}$ is combinatorially equivalent to $\hat{\beta}^*$ if and only if there is a finite sequence of moves

$$(\beta, n) = (\beta_0, n_0) \to (\beta_1, n_1) \to \ldots. \to (\beta_{s-1}, n_{s-1}) \to (\beta_s, n_s) = (\beta^*, n^*)$$

joining $(\beta, n)$ to $(\beta^*, n^*)$, such that for each $0 \leq i < s$, the braid $(\beta_{i+1}, n_{i+1})$ can be obtained from its predecessor $(\beta_i, n_i)$ by applying one of the following moves:

$\mathcal{M}_1$: Replace $\beta_i$ by any other braid in $B(n_i)$, which is conjugate to $\beta_i$. Set $n_{i+1} = n_i$.

$\mathcal{M}_2$: Replace $(\beta_i, n_i)$ by $(\beta_i \sigma_{n_i}^{\pm 1}, n_i + 1)$ ; or, if $\beta_i = \gamma \sigma_{n_i-1}^{\pm 1}$, where the braid word $\gamma$ only involves the generators $\sigma_1, \ldots, \sigma_{n_i-2}$, replace $(\beta_i, n_i)$ by $(\gamma, n_i - 1)$.

The moves $\mathcal{M}_1$ and $\mathcal{M}_2$ are called Markov moves.

The theorem of Markov turns the topological problem of classifying link types in $\mathbb{E}^3$ into an algebraic problem involving the family of braid groups. To be specific: Let $\{(\beta, n) \mid \beta \in B(n)\}$ denote the disjoint union of all the braid groups and define two braids $(\beta, n)$ and $(\beta^*, n^*)$ to be equivalent if one arises from the other by a finite sequence of Markov moves. Then the equivalence classes of braids under this equivalence relation can be identified with the equivalence classes of links in $\mathbb{E}^3$ with respect to combinatorial equivalence.

Unfortunately, although relatively easy to define, Markov's equivalence relation is not very explicit, and very hard to apply directly. For instance, given a link $V$, the problem of finding the braid index for the link, i.e. the smallest number $n$ for which there is a braid $\beta \in B(n)$ with $\hat{\beta} = V$, can be extremely difficult. In general, Markov's theorem is therefore not a very practical method for deciding whether two links are equivalent or not. But the theorem has of course considerable theoretical interest.

Sketch of proof of Theorem 3.1. (Geometric version).

Let $V$ and $V^*$ be combinatorially equivalent closed braids in $\mathbb{E}^3$. Without loss of generality we can – possibly after a series of elementary deformations – assume, that $V$ and $V^*$ have the same positive axis $\ell$ and that they are oriented such that all edges are positive with respect to $\ell$.

By definition of combinatorial equivalence, there is a finite sequence of applicable elementary deformations changing $V$ to $V^*$, say

$$V = V_0 \to V_1 \to \ldots \to V_{s-1} \to V_s = V^* .$$

We know that the heights of $V$ and $V^*$ satisfy $h(V) = h(V^*) = 0$. So, what can happen during the sequence of elementary deformations changing $V$ to $V^*$?

As long as $h(V_i) = h(V_{i+1}) = 0$, the elementary deformation changing $V_i$ to $V_{i+1}$ must be of type $\mathcal{R}$.

Suppose now that negative edges are introduced at stage $i_0$. The elementary deformation $V_{i_0} \to V_{i_0+1}$ may introduce one or two negative edges. Say, that it introduces just one negative edge as pictured in Figure 11.

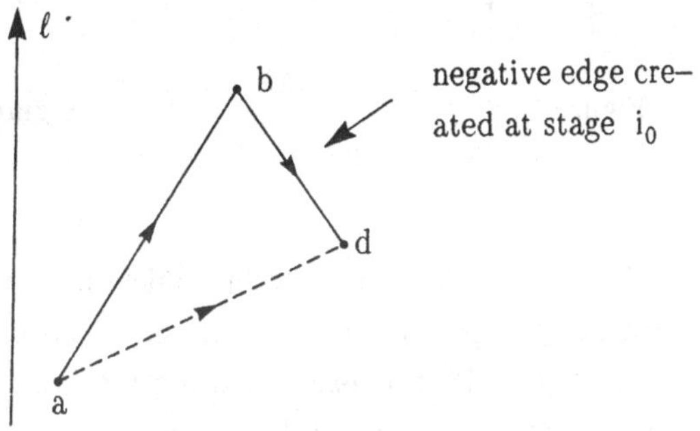


Figure 11

The negative edge introduced at stage $i_0$ must be annihilated by later deformations, since the final result shall be $V^*$ with $h(V^*) = 0$.

One possible way this annihilation can take place is by a sawtooth as pictured in Figure 12. The teeth may be introduced at different stages in later deformations. If the sawtooth contains just one tooth the picture will be as in Figure 13.

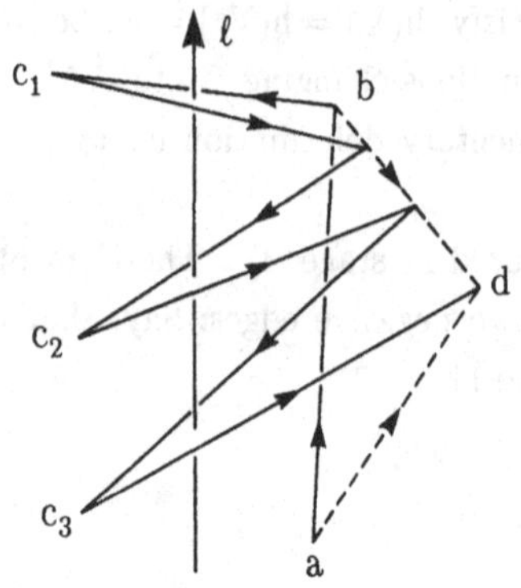

Figure 12

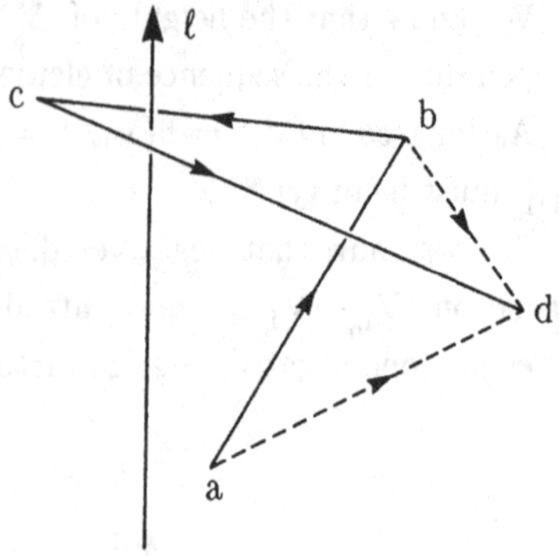

Figure 13

If the picture is as in Figure 13, the combined deformations $\mathscr{E}^{\,b}_{ad}$ and then $\mathscr{E}^{\,c}_{bd}$ is exactly a deformation of type $\mathscr{W}$. By introducing extra deformations along the triangle $\Delta(a, d, b)$ in Figure 12, the general sawtooth operation can be resolved into a sequence of deformations along sawtooth's each with one tooth. After reorganisation, the deformations involved can be collected into a sequence of deformations of type $\mathscr{W}$.

Unfortunately, the annihilation of a negative edge may also take place in other ways. It may in fact be, that it is removed eventually through a sequence of deformations, which at first introduce more and more negative edges. It is a rather delicate combinatorial argument to keep track of the various possibilities, and hence we shall not go further into the proof. Details can be found in Birman's book, pages 48–67.

Sketch of proof of Theorem 3.1 (Algebraic version).

Recall from § 2, that given a closed braid V in $\mathbb{E}^3$ with axis $\ell$, we can recover an open braid $\beta$ representing V by cutting V open along a half–plane P determined by $\ell$. The number of strings in this particular braid representation of V is the number of points in which V intersects P.

It is clear that a deformation of type $\mathscr{R}$ on V will not change the number of strings, whereas a deformation of type $\mathscr{W}$ either increases or lowers the number of strings in the braid representation by 1. Furthermore, the geometry of $\mathscr{W}$ shows that it corresponds to adding or deleting a string in the braid, which is represented by an elementary braid generator $\sigma_i^{\pm 1}$ at either end of the braid representation. Compare with § 1.

Now recall that n–braids are identified with loops in the configuration space $C_n(\mathbb{E}^2)$ based at a particular point $c_0 \in C_n(\mathbb{E}^2)$. When we deform the link corresponding to the braid, the loop defining the braid will move on a free homotopy of loops in $C_n(\mathbb{E}^2)$, i.e. a homotopy that does not necessarily preserve the base point $c_0 \in C_n(\mathbb{E}^2)$. Since a change of base point in $C_n(\mathbb{E}^2)$ corresponds to conjugation in $B(n)$, it follows that a deformation of type $\mathscr{R}$ of the link will correspond to a conjugation in $B(n)$, i.e. a Markov move of type $\mathscr{M}_1$, and that a deformation of type $\mathscr{W}$ of the link will correspond to a Markov move of type $\mathscr{M}_2$, possibly either preceded or followed by a conjugation (Markov move of type $\mathscr{M}_1$).

It is now clear how to translate the geometric version of Markov's theorem into the algebraic version.

## 4. The group of a link.

An important invariant of a link is the fundamental group of the complement of the link. In this section we shall prove a general structure theorem for such link groups.

Let V be a link in euclidean 3–space $\mathbb{E}^3$. By taking the one point compactification of $\mathbb{E}^3$ we obtain the 3–sphere $S^3$, and hence we may also consider a link as an object in $S^3$, when convenient. Since V is 1–dimensional, it follows by simple transversality arguments, that the link complements $\mathbb{E}^3\backslash V$ and $S^3\backslash V$ are connected spaces and that they have the same fundamental group. Hence we get a well defined group

$$G(V) = \pi_1(\mathbb{E}^3\backslash V) = \pi_1(S^3\backslash V) .$$

We need not specify base points, but in computations they are always given implicitly.

The group G(V) is called the group of the link V, or the link group of V.

If V and V' are combinatorially equivalent links in $\mathbb{E}^3$, then they are ambient isotopic, and hence their complements $\mathbb{E}^3\backslash V$ and $\mathbb{E}^3\backslash V'$ are homeomorphic. Consequently, G(V) and G(V') are isomorphic. We list this result as

Lemma 4.1. Combinatorially equivalent links in $\mathbb{E}^3$ have isomorphic link groups.

By the theorem of Alexander (Theorem 2.5), we know that every link in $\mathbb{E}^3$ is combinatorially equivalent to a link $\hat{\beta}$ obtained by closing a braid $\beta \in B(n)$. Hence we only have to worry about closed braids when studying link groups. In Theorem 4.2 below we shall prove a structure theorem for link groups due to Artin [22], 1925. We present an alternative proof due to Joan Birman [1].

First some notation. Let $\mathbb{F}_n$ be the free group on n generators $x_1,\dots,x_n$ for a fixed integer $n \geq 1$. For any set of elements $A_1,\dots,A_n$ in $\mathbb{F}_n$, i.e. each $A_i$ is a word in the generators $x_1,\dots,x_n$, and any permutation $\tau$ of $\{1,\dots,n\}$, we denote by $G(A_1,\dots,A_n ; \tau)$ the abstract group with presentation

generators: $x_1,\dots,x_n$

defining relations: $x_i = A_i\, x_{\tau(i)}\, A_i^{-1}$ for $1 \leq i \leq n$ .

Theorem 4.2. Let $\beta \in B(n)$ be an n–braid, and suppose that the automorphism $\bar{\beta} \in \mathrm{Aut}(\mathbb{F}_n)$, defined by $\beta$, is given by the equations

$$x_i \bar{\beta} = A_i x_{\tau(i)} A_i^{-1} \text{ for } 1 \leq i \leq n ,$$

as in Theorem I.5.1. Then the group $G(\hat{\beta})$ of the link $\hat{\beta}$ in $\mathbb{E}^3$ obtained by closing $\beta$ is isomorphic to the group $G(A_1,...,A_n ; \tau)$, and the elements $A_1,...,A_n$ in $\mathbb{F}_n$ and the permutation $\tau$ of $\{1,...,n\}$ satisfy the identity

$$(*) \qquad (A_1 x_{\tau(1)} A_1^{-1}) \cdot \ldots \cdot (A_n x_{\tau(n)} A_n^{-1}) = x_1 \ldots x_n .$$

Suppose conversely, that the elements $A_1,...,A_n$ in $\mathbb{F}_n$ and the permutation $\tau$ of $\{1,...,n\}$ satisfy the identity (*). Then there exists a braid $\beta \in B(n)$ for which the closed braid $\hat{\beta}$ has link group $G(A_1,...,A_n ; \tau)$.

Proof. We shall use the interpretation of the braid group $B(n)$ given in Theorem I.5.4. Let $D^2$ denote the unit disc in $\mathbb{E}^2$ and $Q_n = \{q_1,...,q_n\}$ a set of $n$ points in the interior of $D^2$. Then we identify the free group $\mathbb{F}_n$ on $n$ generators $x_1,...,x_n$ with the fundamental group of $D^2 \backslash Q_n$, and $B(n)$ with the group of those automorphisms of $\mathbb{F}_n = \pi_1(D^2 \backslash Q_n)$, which are induced by homeomorphisms $h: D^2 \to D^2$ that leave the boundary $S^1 = \partial D^2$ pointwise fixed and permute the points in $Q_n$ among themselves.

Given a braid $\beta \in B(n)$. Let $h: D^2 \to D^2$ be a homeomorphism, which induces the automorphism $\bar{\beta} \in \mathrm{Aut}(\mathbb{F}_n)$ corresponding to $\beta$. According to Theorem I.5.1, the action of $\bar{\beta} = h_*$ is then given by the equations

$$x_i \bar{\beta} = h_*(x_i) = A_i x_{\tau(i)} A_i^{-1} \quad \text{for } 1 \leq i \leq n ,$$

where $A_1,...,A_n$ are certain elements in $\mathbb{F}_n$ and $\tau$ is the permutation of the braid. Furthermore, the identity (*) is satisfied.

We shall now describe a decomposition of the complement of the link $\hat{\beta}$ in the 3-sphere $S^3$, which eventually will allow us to find a presentation for the link group $G(\hat{\beta}) = \pi_1(S^3 \backslash \hat{\beta})$.

The 3-sphere $S^3$ can be constructed as the union of two solid tori glued together along their common boundary. To be specific, we have the following obvious identifications

$$\begin{aligned} S^3 = \partial(D^2 \times D^2) &= D^2 \times \partial D^2 \cup \partial D^2 \times D^2 \\ &= D^2 \times S^1 \cup S^1 \times D^2 = T \cup T' , \end{aligned}$$

where $T = D^2 \times S^1$ and $T' = S^1 \times D^2$ are solid tori glued together along their common boundary torus $S^1 \times S^1$, by identifying the meridian circles of $\partial T = S^1 \times S^1$ with the longitudinal circles of $\partial T' = S^1 \times S^1$, and vice versa.

The link $\hat{\beta}$ is embedded into the solid torus $T$ as follows. By the Alexander trick described in Remark I.5.5, the homeomorphism $h: D^2 \to D^2$ associated with $\beta$ is isotopic to the identity map of $D^2$ along homeomorphisms keeping $S^1 = \partial D^2$ pointwise fixed. Let $H: D^2 \times [0,1] \to D^2 \times [0,1]$ be such an isotopy as defined in Remark I.5.5. Then $\hat{\beta}$ is the link in $T$ which arises from the image of $Q_n \times [0,1]$ in $D^2 \times [0,1]$ under $H$ when the ends of the cylinder $D^2 \times [0,1]$ are glued together. Here and in the following we freely use the identification

$$T = D^2 \times [0,1]/(p, 0) \sim (p, 1) .$$

Using the above identifications, the complement of $\hat{\beta}$ in $S^3$ admits the decomposition

$$S^3 \backslash \hat{\beta} = (T \backslash \hat{\beta}) \cup T' ,$$

where $(T \backslash \hat{\beta}) \cap T' = S^1 \times S^1$.

There is an obvious fibration

$$\pi: T \backslash \hat{\beta} \to S^1$$

of $T\backslash\hat{\beta}$ onto the circle $S^1$ induced from the projection of $T = D^2 \times S^1$ onto the second factor. The fibre of $\pi$ is clearly the space $D^2\backslash Q_n$. Let $\lambda: D^2\backslash Q_n \to T\backslash\hat{\beta}$ be the inclusion of the fibre in the total space defined by the embedding $D^2\backslash Q_n = (D^2\backslash Q_n) \times \{0\} \hookrightarrow T\backslash\hat{\beta}$.

The homotopy sequence for the fibration $\pi$ reduces to the short exact sequence

$$1 \to \pi_1(D^2\backslash Q_n) \xrightarrow{\lambda_*} \pi_1(T\backslash\hat{\beta}) \xrightarrow{\pi_*} \pi_1(S^1) \to 1 ,$$

since all homotopy groups of the spaces involved vanish in dimensions $\geq 2$, and since the fibre $D^2\backslash Q_n$ is connected. This short exact sequence presents $\pi_1(T\backslash\hat{\beta})$ as a group extension of $\pi_1(D^2\backslash Q_n)$ by $\pi_1(S^1)$.

We shall denote the images of the free generators $x_1,\ldots,x_n$ for $\pi_1(D^2\backslash Q_n) = \mathbb{F}_n$ in $\pi_1(T\backslash\hat{\beta})$ under the monomorphism $\lambda_*$ by the same letters.

Let $t \in \pi_1(T\backslash\hat{\beta})$ be the homotopy class represented by the loop in $T\backslash\hat{\beta}$ defined by $q_0 \times [0,1]$, for a point $q_0 \in S^1 = \partial D^2$, and traversed from $q_0 \times \{1\}$ to $q_0 \times \{0\}$. We get a loop since $(q_0, 0) \sim (q_0, 1)$. The loop representing $t \in \pi_1(T\backslash\hat{\beta})$ is a longitude on $\partial T$, and hence it is clear that $\pi_*$ maps $t$ onto the generator $\pi_*(t) \in \pi_1(S^1)$ for the infinite cyclic group $\pi_1(S^1)$.

In order to describe the group $\pi_1(T\backslash\hat{\beta})$ in terms of generators and relations, we now only have to determine the action of $\pi_1(S^1)$ on $\pi_1(D^2\backslash Q_n)$ in the group extension; this amounts to determining the conjugate elements $tx_it^{-1}$ for $1 \leq i \leq n$.

The conjugate element $tx_it^{-1}$ is represented by the loop in $T\backslash\hat{\beta}$, which first follows the longitude $q_0 \times [0,1]$ on $\partial T$ from $(q_0, 1)$ to $(q_0, 0) \sim (q_0, 1)$, then encircles the $i$th string in the closed braid $\hat{\beta}$ once anticlockwise in $(D^2\backslash Q_n) \times \{0\}$, and finally returns to $(q_0, 1)$ along the longitude $q_0 \times [0,1]$. This loop is homotopic to the loop obtained by pulling the loop $x_i$ in $(D^2\backslash Q_n) \times \{0\}$ once around a longitude on $\partial T$ along the gaps in $T$ left by the closed braid $\hat{\beta}$, and hence

$$t\, x_i\, t^{-1} = x_i\beta = h_*(x_i) = A_i\, x_{\tau(i)}\, A_i^{-1} \quad \text{for } 1 \leq i \leq n .$$

The above analysis reveals that the group $\pi_1(T\backslash\hat{\beta})$ admits the presentation

$$\text{generators: } x_1,\ldots,x_n\ ,\ t$$

$$\text{defining relations: } t\,x_i\,t^{-1} = A_i\,x_{\tau(i)}\,A_i^{-1} \quad \text{for } 1 \leq i \leq n\ .$$

To determine $G(\hat{\beta}) = \pi_1(S^3\backslash\hat{\beta})$ we now apply the theorem of van Kampen (see [9]) to the diagram

$$\begin{array}{ccc} & S^3\backslash\hat{\beta} & \\ \nearrow & & \nwarrow \\ T\backslash\hat{\beta} & & T' \\ \nwarrow & & \nearrow \\ & S^1 \times S^1 & \end{array} .$$

Thereby we get the diagram of fundamental groups

$$\begin{array}{ccc} & \pi_1(S^3\backslash\hat{\beta}) & \\ \nearrow & & \nwarrow \\ \pi_1(T\backslash\hat{\beta}) & & \pi_1(T') \\ \nwarrow & & \nearrow \\ & \pi_1(S^1 \times S^1) & \end{array} ,$$

which presents $\pi_1(S^3\backslash\hat{\beta})$ as a free product of $\pi_1(T\backslash\hat{\beta})$ and $\pi_1(T')$ with amalgations defined by $\pi_1(S^1 \times S^1)$ .

The group $\pi_1(S^1 \times S^1)$ is free abelian on two generators M and L , represented respectively by the inclusion of $S^1$ on the first (meridian) and on the second (longitude) factor of $S^1 \times S^1$ .

The generator $M \in \pi_1(S^1 \times S^1)$ is mapped into the element in $\pi_1(T\backslash\hat{\beta})$ represented by a loop encircling all strings in $\hat{\beta}$ just once, and as we have seen before, this is the element $x_1 \ldots x_n$ . The generator $M \in \pi_1(S^1 \times S^1)$ is mapped into the generator for the group $\pi_1(T') = \pi_1(S^1 \times D^2)$ , which is infinite cyclic, and hence the generator for $\pi_1(T')$ shall be identified with the element $x_1 \ldots x_n \in \pi_1(T\backslash\hat{\beta})$ .

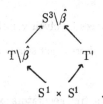

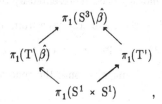

The generator $L \in \pi_1(S^1 \times S^1)$ is mapped into the generator $t \in \pi_1(T\backslash\hat{\beta})$, respectively into $1 \in \pi_1(T')$. This gives the relation $t = 1$.

By the theorem of van Kampen we get therefore the following presentation for the group $G(\hat{\beta}) = \pi_1(S^3\backslash\hat{\beta})$:

generators: $x_1,...,x_n$

defining relations: $x_i = A_i\, x_{\tau(i)}\, A_i^{-1}$ for $1 \leq i \leq n$ .

This proves the first part of Theorem 4.2.

The converse is trivial. If we are given elements $A_1,...,A_n$ in $\mathbb{F}_n$ and a permutation $\tau$ of $\{1,...,n\}$ satisfying the identity (*), then we can define a corresponding braid automorphism $\bar{\beta} \in \mathrm{Aut}(\mathbb{F}_n)$, and thereby a braid $\beta \in B(n)$, according to Theorem I.5.1. The link group of the closed braid $\hat{\beta}$ is then exactly $G(A_1,...,A_n\,;\,\tau)$.

This proves Theorem 4.2.

For a group $G$, which admits a presentation with a set of generators $g_1,...,g_n$ and a set of defining relations $r_1 = r_1(g_1,...,g_n) = 1, \ldots, r_m = r_m(g_1,...,g_n) = 1$, one often uses the notation

$$G = \langle g_1,...,g_n \mid r_1,...,r_m \rangle .$$

The relations $r_1,...,r_m$ are words in the generators $g_1,...,g_n$. We recall that if $F$ is the free group on the generators $g_1,...,g_n$, and $R$ is the normal closure of the elements $r_1,...,r_m$ in $F$, then $G$ is the factor group $G = F/R$. Note also that a defining relation may be written as an equality between words in $F$ and that such an equality can be transformed immediately into an equality of type $r_i = r_i(g_1,...,g_n) = 1$.

Since a braid automorphism of a free group fixes the product of the free generators, any one of the defining relations in the presentation of a link group given in Theorem 4.2 is a consequence of the others. Hence we get

Corollary 4.3. Let $\beta \in B(n)$ be an n–braid, and let $\hat{\beta}$ be the closed braid associated with $\beta$. Then the link group $G(\hat{\beta})$ of $\hat{\beta}$ admits the presentation

$$G(\hat{\beta}) = \langle x_1,\dots,x_n \mid x_1 = x_1\bar{\beta},\dots, x_{n-1} = x_{n-1}\bar{\beta} \rangle ,$$

where $x_1,\dots,x_n$ are generators for the free group $\mathbb{F}_n$, and $\bar{\beta}$ is the automorphism of $\mathbb{F}_n$ induced by $\beta$.

5. Plane projections and braid representations of links.

Let V be a link in $\mathbb{E}^3$ with axis $\ell$. Suppose that both V and $\ell$ are oriented, and that V is in general position with respect to $\ell$. Choose a plane $\pi$ in $\mathbb{E}^3$ orthogonal to $\ell$. The orientation of $\ell$ induces an orientation of the plane $\pi$ – anticlockwise rotation with respect to $\ell$. The point where $\ell$ intersects $\pi$ is marked by $\odot$, thereby also indicating the orientation of $\pi$. If we project V orthogonally onto $\pi$, we get a picture like Figure 14, which is the projection of 3 pairwise interlocking rings in $\mathbb{E}^3$.

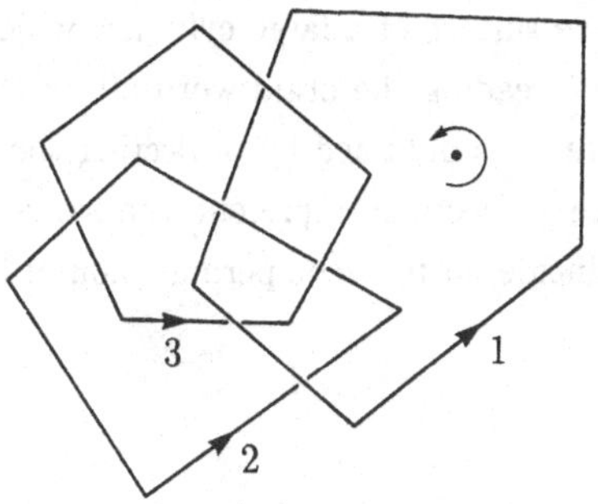


Figure 14

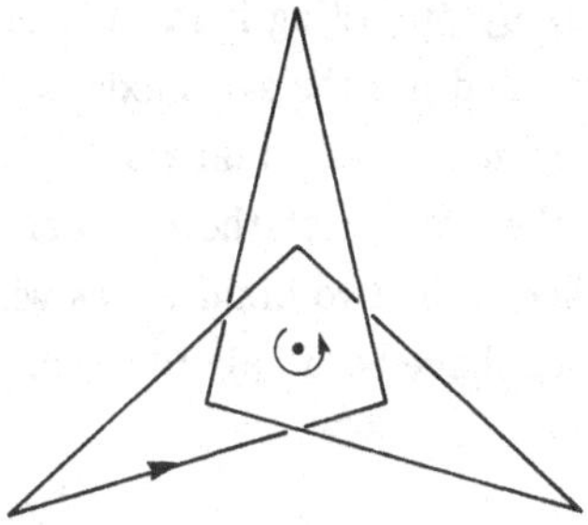

Figure 15

We have indicated over and under crossings of components in the projection of the link. After a small perturbation of the link – a sequence of elementary deformations – we may always assume that all crossings are transversal and that no vertice of the link projects onto a multiple point. Henceforward we shall always assume that our links have been so manufactured.

By Alexander's theorem we know that every link is equivalent to a closed braid. In the projections of links, the closed braids show up as those links, for which the projections of the components all encircle the point $\odot$. The projection in Figure 14 is not the projection of a closed braid, but a parallel translation will make it into one. The projection in Figure 15 is the projection of a closed braid, in fact of the right handed trefoil knot to which we shall return in Example 5.1.

If the projection onto $\pi$ is that of a closed braid V in $\mathbb{E}^3$, one can read off, from the projection, a braid word representing the closed braid as follows: Rotate a half–line in $\pi$ originating from $\odot$ one full turn anticlockwise with respect to the orientation of $\pi$. In an initial position we assign numbers 1,...,n to the points of intersection between the half–line and the projection of the link V. Now, as the half–line is turned around, the points of intersection with the projection of the link will move on the half–line. Cutting open along the initial position of the half–line and bending the family of half–lines into parallel position provides us with the projection of a braid. Taking into account over and under crossings of strings, a braid word representing the link can be read off. See Figure 16.

The braid word representing the closed braid V obtained by the above procedure is in general a different braid word from that used to construct a link from a braid in §1. The braid word behind the construction of a closed braid V in §1 can be recovered from the link by projecting it radially onto the surface of a large cylinder, which contains the link and has the same axis, and then reading the braid word off on the surface of the cylinder. The braid word in Figure 16 is obtained by projecting the closed braid onto the bottom of the cylinder. Both projections represent the same closed braid, and hence the two braid words will coincide up to cyclic permutation of letters, after the words have been fully reduced.

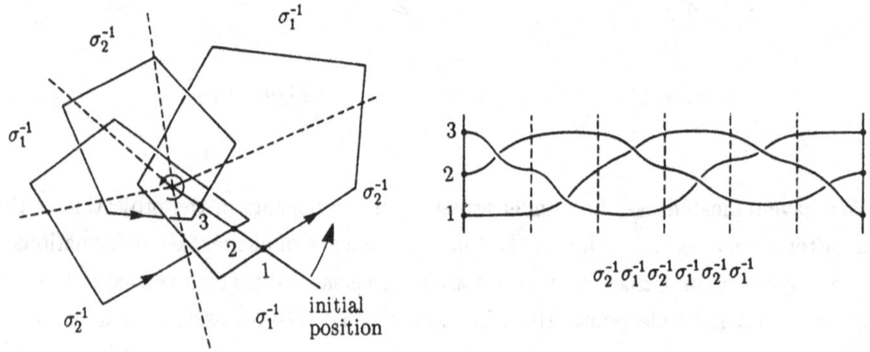


Figure 16

Example 5.1. In Figures 17 and 18 we have pictured projections of the most common knot of all, the trefoil knot. There is a right handed and a left handed trefoil knot, corresponding to the knot tied by a right handed, respectively a left handed, person, when he ties a knot on a piece of string and closes the ends of the string.

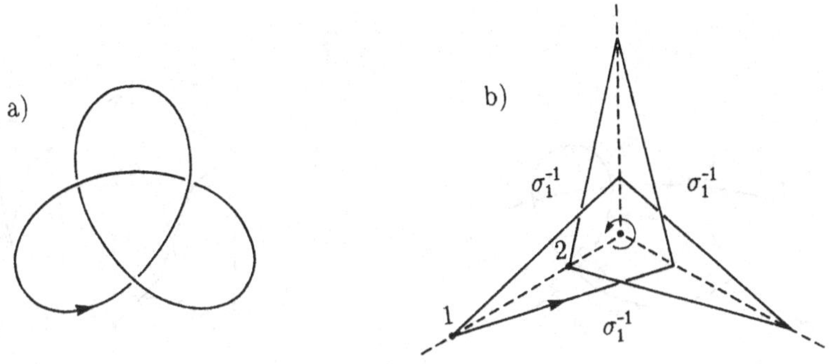


Right handed trefoil knot
Figure 17

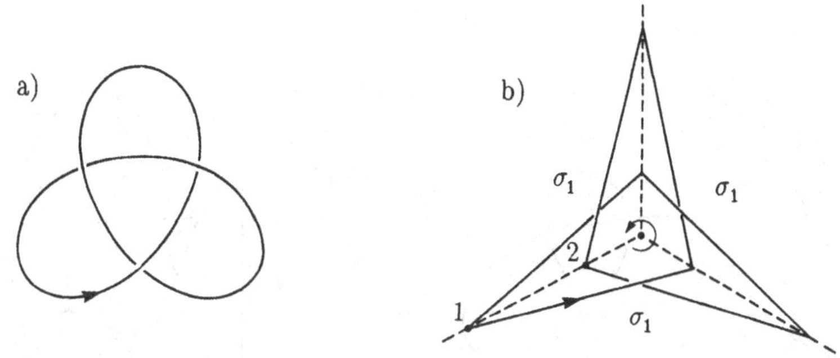


Left handed trefoil knot
Figure 18

The pictures under a) are the standard pictures of the knots, while the pictures under b) are suitable for finding braid representations of the knots.

From the projections in the Figures 17b and 18b, we read off that the right handed trefoil knot is represented by the braid word $\sigma_1^{-3}$ , and the left handed trefoil knot by the braid word $\sigma_1^3$ , both braids with 2 strings.

Example 5.2. Two other common knots, the granny knot and the square knot, are pictured by their projections in Figure 19 and 20. The square knot is the tricky knot which requires thought to make – you take a piece of string and make a right handed knot followed by a left handed knot and ties the ends of the string together. The granny knot is made by making two equal handed knots on the piece of string and then tieing the ends together.

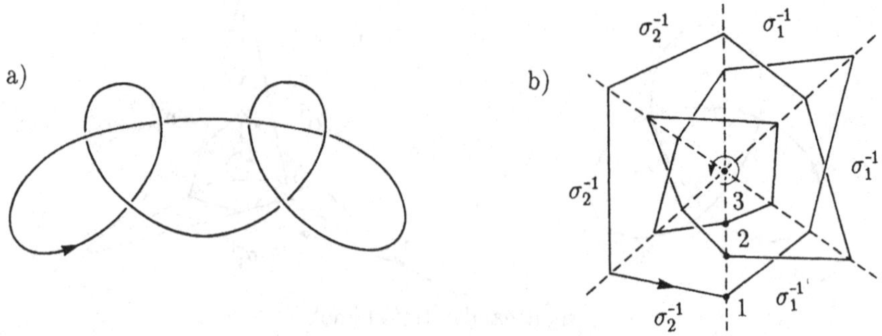


Granny knot
Figure 19

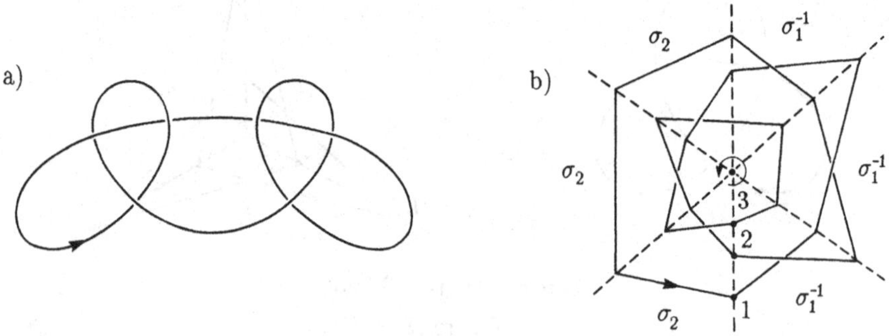


Square knot
Figure 20

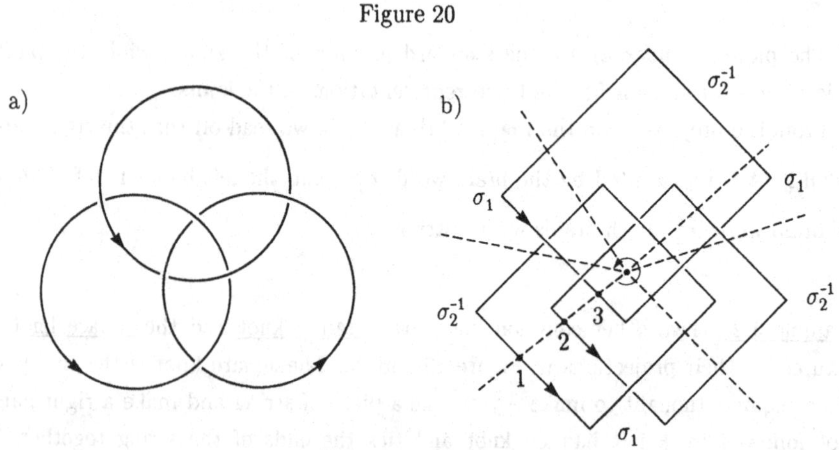


Borromean rings
Figure 21

Again we present two projections of each of the knots. It is clear that we get equivalent knots in case a) and case b) since in both cases we are making two single knots of the appropriate types on a piece of string before tieing the ends together. It is however also quite instructive to convince yourself that they are actually equivalent by trying it out puzzling with a closed string.

From the projections in Figures 19b and 20b, we read off that the granny knot is represented by the braid word $\sigma_1^{-3}\sigma_2^{-3}$ , and the square knot by the braid word $\sigma_1^{-3}\sigma_2^{3}$ , both braids with 3 strings.

Example 5.3. In Figure 21 we have pictured the projection of the so-called Borromean rings. It is a link with 3 components with the property that if any one of the components is removed then it falls completely apart. In other words: Any two of the three components form the trivial link of two components.

From the projection in Figure 21b we read off that the Borromean rings is represented by the braid word $(\sigma_1\sigma_2^{-1})^3$ . This is actually the link made by a suitable piece of a girl's plait.

From a braid representation of a link $V$ , one can compute the link group $G(V)$ of $V$ using Corollary 4.3.

As an example we compute the link group of the right handed trefoil knot. According to Example 5.1, this amounts to computing $G(\hat{\sigma}_1^{-3})$ , since the 2-string braid $\sigma_1^{-3}$ represents the knot in question. By Corollary 4.3 we have

$$\begin{aligned} G(\hat{\sigma}_1^{-3}) &= \langle x_1, x_2 \mid x_1 = x_1\, \overline{\sigma}_1^{-3} \rangle \\ &= \langle x_1, x_2 \mid x_1 = x_2\, \overline{\sigma}_1^{-2} \rangle \\ &= \langle x_1, x_2 \mid x_1 = (x_2^{-1}\, x_1\, x_2)\, \overline{\sigma}_1^{-1} \rangle \\ &= \langle x_1, x_2 \mid x_1 = (x_2^{-1}\, x_1^{-1}\, x_2)\, x_2\, (x_2^{-1}\, x_1\, x_2) \rangle \\ &= \langle x_1, x_2 \mid x_1\, x_2\, x_1 = x_2\, x_1\, x_2 \rangle \ . \end{aligned}$$

This is exactly the presentation of the braid group on 3 strings, compare with Theorem I.1.5. Therefore, the right handed trefoil knot has the link group

$$G(\hat{\sigma}_1^{-3}) = B(3) \ .$$

The left handed trefoil knot has also link group $B(3)$, since the complements in $\mathbb{E}^3$ of the right handed and the left handed trefoil knots are homeomorphic by an orientation reversing homeomorphism, namely reflection in a plane.

The trivial knot, or the unknot, is a standard embedded circle in $\mathbb{E}^3$. It is represented by the trivial braid $\varepsilon \in B(1)$ on 1 string. According to Corollary 4.3, the trivial knot has therefore link group

$$G(\hat{\varepsilon}) = \langle x_1 \rangle = \mathbb{Z} \ ,$$

which is the free (abelian) group on 1 generator. This is also easy to prove without using Corollary 4.3.

It follows by Lemma 4.1 that the trefoil knot and the trivial knot are not equivalent. In other words: The trefoil knot cannot be unknotted without untieing it. So, non-trivial knots do exist!

## EXERCISES

1. Prove that combinatorially equivalent links V and V' in $\mathbb{E}^3$ are ambient isotopic.

2. Prove that the links obtained by closing the 4–braid $(\sigma_1\sigma_2\sigma_3)^2$ and the 2–braid $\sigma_1^4$ are combinatorially equivalent.

3. Apply the procedure in the proof of Alexander's theorem to the standard projection of the granny knot to obtain the projection of a closed braid. Find a braid representative for this closed braid and prove that it is equivalent to the closed braid $\sigma_1^{-3}\sigma_2^{-3}$.

Try also the square knot.

4. Find presentations of the link groups of the granny knot, the square knot and the Borromean rings.

5. A link in $\mathbb{E}^3$, which is combinatorially equivalent to a link consisting of $m \geq 1$ disjoint circles in a plane, is called a <u>trivial link</u> of m components. Prove that all trivial links of m components are combinatorially equivalent and have the free group on m generators as link group.

6. Consider the link consisting of two interlocking circles in $\mathbb{E}^3$. Find a braid representative for this link and prove that its link group is the free abelian group on 2 generators. Conclude that the link is not trivial.

7. Divide a standard torus (the surface of a doughnut) in $b \geq 2$ equal pipes ("cylinders"), and mark on each of the b separating meridian circles $a \geq 2$ points. Orient the meridian circles coherently and also orient the torus. Thereby every pipe gets a negative and a positive boundary circle. Number the points on each separating meridian circle 1,...,a in accordance with the orientation of the circle. On each pipe connect point number i on the negative boundary circle by a right "helicoid" to point number $i+1$ (counted modulo a) on the positive boundary circle. Thereby we get a system of embedded circles on the torus, or in other words, a link V(a, b) in $\mathbb{E}^3$ which lies on the torus.

(i) Suppose that a and b are relatively prime. Prove that V(a, b) is a knot. This knot is called the <u>torus knot</u> of type (a, b).

(ii) Prove that the torus knot $V(a, b)$ for $a$ and $b$ relatively prime, traverses the longitude on the torus $a$ times and the meridian $b$ times.

(iii) Draw plane projections of the torus knots $V(2, 3)$ and $V(2, 5)$. Notice that $V(2, 3)$ is the trefoil knot. The knot $V(2, 5)$ is called Solomon's knot.

(iv) Find a braid representative for Solomon's knot $V(2, 5)$ and compute its link group.

(v) Prove that for arbitrary numbers $a \geq 2$ and $b \geq 2$, the number of components $m$ of $V(a, b)$ is the greatest common divisor of $a$ and $b$.

Chapter III

# POLYNOMIAL COVERING MAPS

In his study of the geometry of (germs of) zero sets for holomorphic functions of several complex variables, Weierstrass proved around 1880 his famous Preparation Theorem, which reduces this study to parametrized families of complex polynomials with holomorphic coefficients. By adapting these concepts to the continuous setting, one is led to an interesting class of finite covering maps, introduced in [33] and [34] under the name polynomial covering maps. From an algebraic point of view, polynomial covering maps correspond to conjugacy classes of homomorphisms of fundamental groups into the Artin braid groups. The theory of this special class of finite covering maps is thereby closely related to the theory of braid groups.

The first section of Chapter III contains the basic definitions of Weierstrass polynomials and the covering maps associated with them. The exploration of these concepts will be our main concern in Chapters III and IV.

In Chapter III we present three characterizations of the class of polynomial covering maps. First we prove the existence of a canonical n–fold polynomial covering map onto the complement $B^n = \mathbb{C}^n \setminus \Delta$ of the discriminant set $\Delta$ in complex n–space $\mathbb{C}^n$, from which every n–fold polynomial covering map is induced by a pull–back construction. The second characterization is an embedding criterion, according to which a finite covering map onto a space X is equivalent to a polynomial covering map if and only if it embeds (fibrewise) into the trivial complex line bundle over X. The third characterization describes the n–fold polynomial covering maps onto X in terms of conjugacy classes of homomorphisms of the fundamental group of X into the Artin braid group B(n) on n strings. This connection is established via characteristic homomorphisms for n–fold covering maps, and follows from the fact that $B^n = \mathbb{C}^n \setminus \Delta$ is an Eilenberg–MacLane space of type (B(n), 1). Based on this very fruitful point of view we obtain a complete algebraic classification of the polynomial covering maps.

Motivated by the embedding criterion for polynomial covering maps, several authors have studied embedding problems for finite covering maps into arbitrary bundles. In the final section of Chapter III, we present a new general embedding theorem for finite covering maps into bundles of manifolds.

## 1. Weierstrass polynomials and the finite covering maps associated with them.

Throughout this chapter, $X$ denotes a connected and locally pathwise connected topological space. When necessary, we shall further assume that $X$ has the homotopy type of a CW–complex and that $x_0 \in X$ is a nondegenerate base point in $X$.

Let $\mathbb{C}$ denote the complex number space. We shall freely identify $\mathbb{C}$ with the euclidean plane $\mathbb{E}^2$.

Definition 1.1. A Weierstrass polynomial of degree $n \geq 1$ over $X$ is a polynomial map $P: X \times \mathbb{C} \to \mathbb{C}$ of the form

$$P(x, z) = z^n + \sum_{i=1}^{n} a_i(x)\, z^{n-i}$$
$$= z^n + a_1(x)\, z^{n-1} + \ldots + a_{n-1}(x)\, z + a_n(x) ,$$

where $a_1,\ldots,a_n : X \to \mathbb{C}$ are continuous, complex valued functions on $X$, and $z \in \mathbb{C}$ is a complex variable.

We say that $P(x, z)$ is a simple (or separable) Weierstrass polynomial, if it has no multiple roots for any $x \in X$.

Let $C(X)$ denote the ring of continuous, complex valued functions $f: X \to \mathbb{C}$ on $X$ with its usual pointwise defined algebraic structure, and let $C(X)[z]$ denote the polynomial ring in one complex variable $z$ over $C(X)$. A Weierstrass polynomial is then a monic (leading coefficient 1) polynomial in $C(X)[z]$.

The origin to the name Weierstrass polynomial lies in work of Weierstrass around 1880 culminating in the famous Weierstrass Preparation Theorem. This theorem reduces the study of the geometry (topology) of the zero set for a (germ of a) holomorphic function in $k + 1$ complex variables with nontrivial Taylor expansion to the study of the zero set for a complex polynomial – of the type we now call a Weierstrass polynomial – with holomorphic functions in $k$ complex variables as coefficients.

Consider an arbitrary Weierstrass polynomial $P(x, z)$ of degree $n \geq 1$ over $X$. For each point $x \in X$, the Weierstrass polynomial $P(x, z)$ defines a single complex polynomial of degree $n \geq 1$. We may therefore also think of a Weierstrass polynomial as a continuous family of complex polynomials parametrized by $X$.

Suppose now that $P(x, z)$ is simple. Then each polynomial in the family has $n$ distinct complex roots. If we plot these roots over the corresponding points in the parameter space $X$, we get an n–fold covering map. To be precise we proceed as follows. Let

$$E = \{(x, z) \in X \times \mathbb{C} \mid P(x, z) = 0\}$$

be the set of all zeros for $P(x, z)$ for all $x \in X$, topologized as a subset of the product space $X \times \mathbb{C}$, and let

$$\pi: E \to X$$

be the projection map onto the first factor. The inclusion of $E$ into $X \times \mathbb{C}$ exhibits the map $\pi: E \to X$ as embedded into the trivial complex line bundle $\mathrm{proj}_1: X \times \mathbb{C} \to X$ over $X$,

$$\begin{array}{ccc} E & \hookrightarrow & X \times \mathbb{C} \\ & {\scriptstyle \pi} \searrow \quad \swarrow {\scriptstyle \mathrm{proj}_1} & \\ & X & \end{array} .$$

For each point $x \in X$, the inverse image $\pi^{-1}(x)$ contains $n$ points, namely the roots of the polynomial which corresponds to $x$. We shall prove in § 3 that $\pi: E \to X$ is actually an n–fold covering map. This leads to the following

<u>Definition 1.2.</u> Let $P(x, z)$ be a simple Weierstrass polynomial of degree $n \geq 1$ over $X$. Then $\pi: E \to X$ is called the n–fold <u>polynomial covering map</u>, and $E$ the n–fold <u>polynomial covering space</u> over $X$, associated with $P(x, z)$.

A geometric picture of the polynomial covering map $\pi: E \to X$ is given in Figure 1.

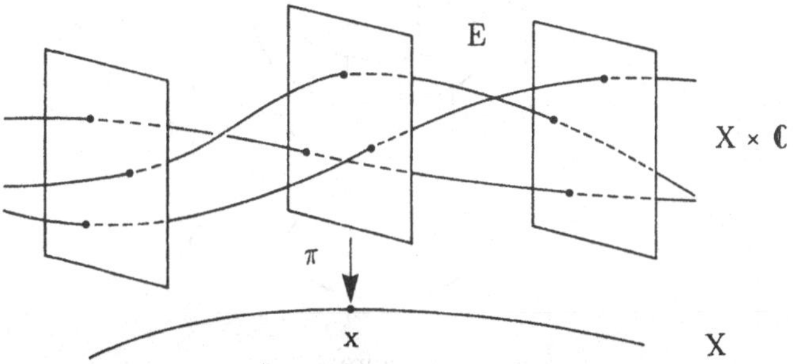


Figure 1

<u>Example 1.3.</u> For an arbitary topological space $X$, we can define a Weierstrass polynomial of degree $n \geq 1$ over $X$ by

$$P(x, z) = (z-1)(z-2)\cdot\ldots\cdot(z-n) .$$

Clearly $P(x, z)$ is simple. The polynomial covering map $\pi\colon E \to X$ associated with $P(x, z)$ is obviously the <u>trivial</u> n–fold covering map $\mathrm{proj}_1\colon X \times \{1,\ldots,n\} \to X$ onto $X$.

<u>Example 1.4.</u> Let $S^1$ be the unit circle, considered as the set of complex numbers $x \in \mathbb{C}$ of modulus $|x| = 1$. For $n \geq 2$, we define the Weierstrass polynomial

$$P(x, z) = z^n - x \ , \quad x \in S^1 , z \in \mathbb{C} \ ,$$

over $S^1$. Clearly $P(x, z)$ is simple, and hence we get an n–fold polynomial covering map $\pi\colon E \to S^1$. This covering map is equivalent to the n–fold covering map $p_n\colon S^1 \to S^1$ defined by $p_n(z) = z^n$. A homeomorphism $h\colon S^1 \to E$ commuting with projections onto $S^1$ can be defined by $h(z) = (z^n, z)$.

In Figure 2 we have pictured the case $n = 2$. Here we get the nontrivial double covering of the circle, which can also be defined by projecting the boundary circle of the Möbius band onto the central circle.

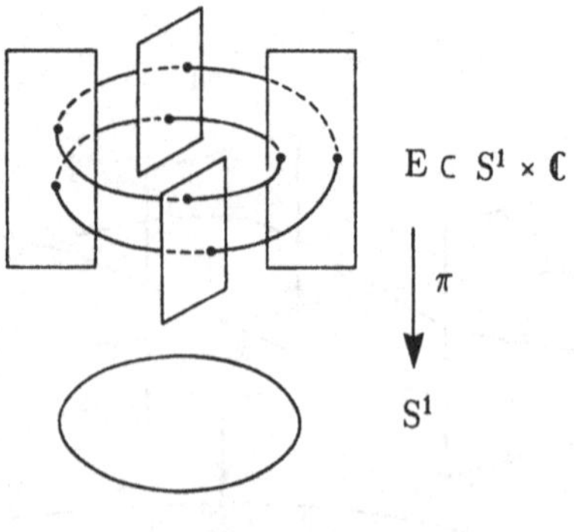


Figure 2

We shall give more examples as we proceed.

The basic problem to be considered is the following

Problem. Given a connected and locally pathwise connected topological space $X$. Describe the equivalence classes of the finite covering maps onto $X$ that contain polynomial covering maps. In particular, find criteria that a finite covering map is equivalent to a polynomial covering map.

We recall that two covering maps onto $X$, $\pi_i\colon E_i \to X$, $i = 1,2$, are said to be equivalent, if there exists a homeomorphism $h\colon E_1 \to E_2$, which commutes with projections onto $X$,

$$\begin{array}{ccc} E_1 & \xrightarrow{h} & E_2 \\ & {}_{\pi_1}\searrow \quad \swarrow_{\pi_2} & \\ & X & \end{array} .$$

An n–fold covering map $\pi\colon E \to X$, which is equivalent to the trivial n–fold covering map $\mathrm{proj}_1\colon X \times \{1,\ldots,n\} \to X$ will itself be called trivial.

As we shall see, there are many finite covering maps, which are not equivalent to polynomial covering maps. In fact, there are well known spaces which do not admit any nontrivial polynomial covering spaces at all.

## 2. The canonical n–fold polynomial covering map.

Let $\mathbf{C}^n$ denote the n–dimensional complex number space, and let $\Delta \subset \mathbf{C}^n$ be the subset consisting of those complex n–tuples $(a_1,\ldots,a_n) \in \mathbf{C}^n$ for which the complex polynomial $P(z) = z^n + \sum_{i=1}^{n} a_i z^{n-i}$ has at least one multiple root. The subset $\Delta$ is called the discriminant set in $\mathbf{C}^n$.

We make now a small detour to describe the discriminant set further.

Let $s_1(x_1,\ldots,x_n),\ldots,s_n(x_1,\ldots,x_n)$ denote the n elementary symmetric polynomials in n variables $x_1,\ldots,x_n$. Recall that these polynomials are defined by the polynomial identity

$$\prod_{i=1}^{n} (x - x_i) = x^n + \sum_{i=1}^{n} (-1)^i s_i(x_1,\ldots,x_n) x^{n-i} .$$

We get

$$\begin{aligned} s_1(x_1,\ldots,x_n) &= x_1+\ldots+x_n \\ s_2(x_1,\ldots,x_n) &= \sum_{1\leq i<j\leq n} x_i x_j \\ &\vdots \\ s_n(x_1,\ldots,x_n) &= x_1\cdot\ldots\cdot x_n . \end{aligned}$$

It is well known [14] that any symmetric polynomial in n variables can be written as a unique polynomial in the elementary symmetric polynomials. Hence there is a unique polynomial in n variables $\delta(a_1,\ldots,a_n)$ for which

$$\delta(-s_1(x_1,\ldots,x_n),\ldots,(-1)^i s_i(x_1,\ldots,x_n),\ldots,(-1)^n s_n(x_1,\ldots,x_n)) = \prod_{1\leq i<j\leq n} (x_i - x_j)^2 .$$

The polynomial $\delta(a_1,\ldots,a_n)$ is called the discriminant polynomial. For a concrete complex polynomial $P(z) = z^n + \sum_{i=1}^{n} a_i z^{n-i}$, the complex number $\delta(a_1,\ldots,a_n)$ is called the discriminant of the polynomial. For $n = 2$, we get the well known formula $\delta(a_1, a_2) = a_1^2 - 4a_2$.

By the fundamental theorem of algebra, a complex polynomial $P(z) = z^n + \sum_{i=1}^{n} a_i z^{n-i}$ admits a factorization into linear factors,

$$P(z) = z^n + \sum_{i=1}^{n} a_i z^{n-i} = \prod_{i=1}^{n} (z - \alpha_i) .$$

The factorization is unique up to a permutation of the factors, and the numbers $\alpha_1,\dots,\alpha_n$ are the n complex roots of $P(z)$, counted with multiplicity. We note that

$$a_i = (-1)^i s_i(\alpha_1,\dots,\alpha_n)$$

and hence that

$$\delta(a_1,\dots,a_n) = \prod_{1\le i<j\le n} (\alpha_i - \alpha_j)^2 .$$

From this formula it follows immediately that $(a_1,\dots,a_n) \in \Delta$ if and only if $\delta(a_1,\dots,a_n) = 0$. In other words: A complex polynomial has multiple roots if and only if its discriminant vanishes, or, from our point of view, the discriminant set $\Delta$ in $\mathbb{C}^n$ is exactly the set of zeros for the discriminant polynomial $\delta(a_1,\dots,a_n)$. This proves

Lemma 2.1. The discriminant set $\Delta$ is an algebraic variety of complex codimension 1 in $\mathbb{C}^n$.

After this detour we shall proceed to define a canonical n–fold polynomial covering space over the complement of the discriminant set.

Let $B^n = \mathbb{C}^n \backslash \Delta$ denote the complement of the discriminant set $\Delta$ in $\mathbb{C}^n$. By Lemma 2.1 it follows in particular, that $B^n$ is a connected (and locally pathwise connected) open subset of $\mathbb{C}^n$.

Over $B^n$ we have a canonical Weierstrass polynomial, namely

$$P^n((a_1,\dots,a_n), z) = z^n + \sum_{i=1}^{n} a_i z^{n-i} ,$$

where $(a_1,\dots,a_n) \in B^n$ and $z \in \mathbb{C}$. By definition of $\Delta$, it is clear that $P^n$ is a simple Weierstrass polynomial over $B^n$. The associated n–fold polynomial covering map $\pi^n\colon E^n \to B^n$ is called the canonical n–fold polynomial covering map. We have the diagram

$$\begin{array}{ccc} E^n & \hookrightarrow & B^n \times \mathbb{C} \\ & {\scriptstyle \pi^n}\searrow \quad \swarrow {\scriptstyle \mathrm{proj}_1} & \\ & B^n & \end{array} .$$

Lemma 2.2. The canonical n–fold polynomial covering map $\pi^n \colon E^n \to B^n$ is an n–fold covering map.

Proof. Let $((a_1^\circ,\dots,a_n^\circ), z_0) \in E^n$. By definition $P^n((a_1^\circ,\dots,a_n^\circ), z_0) = 0$. Since $z_0$ is a simple root in the polynomial $P^n((a_1^\circ,\dots,a_n^\circ), z)$, we have

$$\frac{\partial P^n}{\partial z}((a_1^\circ,\dots,a_n^\circ), z_0) \neq 0 .$$

By the Implicit Function Theorem for holomorphic functions, there exists therefore a holomorphic function $\varphi \colon U \to \mathbb{C}$ defined on an open neighbourhood $U \subset B^n$ of $(a_1^\circ,\dots,a_n^\circ) \in B^n$, such that $P^n((a_1,\dots,a_n), \varphi(a_1,\dots,a_n)) = 0$ for $(a_1,\dots,a_n) \in U$ and $\varphi(a_1^\circ,\dots,a_n^\circ) = z_0$. The function $\varphi$ defines a local section for $\pi^n$ over $U$, namely

$$(a_1,\dots,a_n) \to ((a_1,\dots,a_n), \varphi(a_1,\dots,a_n)) .$$

Using these local sections it is easy to prove, that $\pi^n$ is an n–fold covering map. This proves Lemma 2.2.

There is also another approach to the canonical n–fold polynomial covering map. This approach uses configuration spaces and will be described now.

Recall that

$$F_n(\mathbb{C}) = \{(z_1,\dots,z_n) \in \mathbb{C}^n \mid z_i \neq z_j, i \neq j\}$$

denotes the space of n–tuples of pairwise distinct complex numbers. The symmetric group on n elements $\Sigma_n$ acts freely on $F_n(\mathbb{C})$ by permutation of coordinates,

$$F_n(\mathbb{C}) \times \Sigma_n \to F_n(\mathbb{C}) ,$$

$$((z_1,\dots,z_n), \sigma) \mapsto (z_{\sigma(1)},\dots,z_{\sigma(n)})$$

with orbit space,

$$C_n(\mathbb{C}) = {}^{F_n(\mathbb{C})}\!\big/_{\Sigma_n} \ .$$

Projection onto the orbit space defines the principal $\Sigma_n$ – bundle,

$$p_n: F_n(\mathbb{C}) \to C_n(\mathbb{C}) \ .$$

As we know, the space $F_n(\mathbb{C})$ , respectively $C_n(\mathbb{C})$ , is the configuration space for a set of n ordered, respectively unordered, pairwise distinct points in the complex plane $\mathbb{C}$ .

If we let $\Sigma_n$ act to the left on the set of numbers $\{1,...,n\}$ ,

$$\Sigma_n \times \{1,...,n\} \to \{1,...,n\} \ ,$$
$$(\sigma, i) \mapsto \sigma(i)$$

then we can form the associated bundle

$$\overline{p}_n: F_n(\mathbb{C}) \times_{\Sigma_n} \{1,...,n\} \to C_n(\mathbb{C}) \ ,$$

to the principal $\Sigma_n$–bundle $p_n$ . Note that $\overline{p}_n$ is, in particular, an n–fold covering map.

By the fundamental theorem of algebra (for analysists) there are canonical homeomorphisms

$$\Phi: C_n(\mathbb{C}) \to B^n$$

and

$$\overline{\Phi}: F_n(\mathbb{C}) \times_{\Sigma_n} \{1,...,n\} \to E^n$$

defined by

$$\Phi([\alpha_1,...\alpha_n]) = (a_1,...,a_n)$$

and

$$\overline{\Phi}([(\alpha_1,...,\alpha_n), i]) = ((a_1,...,a_n), \alpha_i) \ ,$$

when

$$z^n + \sum_{i=1}^{n} a_i z^{n-i} = \prod_{i=1}^{n} (z - \alpha_i) \ ,$$

and [ ] indicates that an equivalence class has been taken.

The diagram

$$\begin{array}{ccc} F_n(\mathbb{C}) \times_{\Sigma_n} \{1,\dots,n\} & \xrightarrow{\;\;\overline{\Phi}\;\;} & E^n \\ {\scriptstyle \overline{p}_n}\big\downarrow & & \big\downarrow{\scriptstyle \pi^n} \\ C_n(\mathbb{C}) & \xrightarrow[\;\;\Phi\;\;]{} & B^n \end{array}$$

is clearly commutative, and it defines an identification of the canonical n–fold polynomial covering map $\pi^n \colon E^n \to B^n$ and the $\Sigma_n$ – bundle (in the sense of Steenrod [13]) $\overline{p}_n \colon F_n(\mathbb{C}) \times_{\Sigma_n} \{1,\dots,n\} \to C_n(\mathbb{C})$ .

## 3. Geometric characterizations of polynomial covering maps.

In this section we shall present two geometric characterizations of polynomial covering maps. The first is a pull–back criterion and the second an embedding criterion.

Let $\pi\colon E \to X$ be an n–fold polynomial covering map associated with the simple Weierstrass polynomial

$$P(x, z) = z^n + \sum_{i=1}^{n} a_i(x)\, z^{n-i}$$

over $X$. By definition of $B^n$ as the complement of the discriminant set $\Delta \subset \mathbb{C}^n$, it is clear that the coefficient functions in $P(x, z)$ define a map

$$a = (a_1,\ldots,a_n) : X \to B^n .$$

For obvious reasons we call a <u>the coefficient map</u> for the Weierstrass polynomial $P(x, z)$ and the polynomial covering map $\pi\colon E \to X$.

In the following we shall often denote the Weierstrass polynomial and the polynomial covering map belonging to the coefficient map $a$ by the symbols $P_a(x, z)$, respectively $\pi_a\colon E_a \to X$.

Let

$$\begin{array}{ccc} a^*(E^n) & \xrightarrow{\ a^*\ } & E^n \\ {\scriptstyle (\pi^n)^*}\downarrow & & \downarrow{\scriptstyle \pi^n} \\ X & \xrightarrow[\ a\ ]{} & B^n \end{array}$$

denote the pull–back of the canonical n–fold polynomial covering map along the coefficient map $a$. Then there is a canonical equivalence of covering maps onto $X$,

$$\begin{array}{ccc} E_a & \xrightarrow{\ h\ } & a^*(E^n) \\ & {\scriptstyle \pi_a}\searrow \quad \swarrow{\scriptstyle (\pi^n)^*} & \\ & X & \end{array} ,$$

defined by $h(x, z) = (x, (a_1(x),\ldots,a_n(x), z))$ for $(x, z) \in E_a$.

We list this observation as

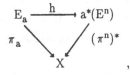

Theorem 3.1. Any n–fold polynomial covering map $\pi_a: E_a \to X$ is equivalent to the pull–back of the canonical n–fold polynomial covering map along its coefficient map $a: X \to B^n$.

Since we know by Lemma 2.2 that the canonical n–fold polynomial covering map is, in particular, an ordinary covering map we get immediately the following corollary from Theorem 3.1.

Corollary 3.2. Every polynomial covering map is an ordinary covering map.

Using a well known property of pull–backs of a bundle we get also

Corollary 3.3. Let $\pi_a: E_a \to X$ and $\pi_b: E_b \to X$ be polynomial covering maps with coefficient maps $a: X \to B^n$ and $b: X \to B^n$, respectively. Then $\pi_a$ and $\pi_b$ are equivalent as covering maps if a and b are homotopic as maps of X into $B^n$.

Corollary 3.3 can also be proved directly without reference to the pull–back construction. In fact, an equivalence between $\pi_a$ and $\pi_b$ can be constructed using the unique path lifting in a polynomial covering map $\pi_H: E_H \to X \times [0,1]$ with coefficient map $H: X \times [0,1] \to B^n$, where H is a homotopy such that $H \mid X \times \{0\} = a$ and $H \mid X \times \{1\} = b$.

By a slight reformulation of Theorem 3.1 we get also the first geometric recognition principle for polynomial covering maps, namely

Theorem 3.4. The equivalence class of an n–fold covering map $\pi: E \to X$ contains a polynomial covering map if and only if it contains a pull–back of the canonical n–fold polynomial covering map $\pi^n: E^n \to B^n$ along a map $a: X \to B^n$.

The second geometric recognition principle for polynomial covering maps is an embedding criterion.

Definition 3.5. We say that a finite covering map $\pi\colon E \to X$ can be embedded into the trivial complex line bundle over $X$, if there is a continuous map (called an embedding) $h\colon E \to X \times \mathbb{C}$, which maps $E$ homeomorphically onto its image $h(E)$ in $X \times \mathbb{C}$ and commutes with projections onto $X$,

$$\begin{array}{ccc} E & \xrightarrow{\;h\;} & X \times \mathbb{C} \\ & {\scriptstyle\pi}\searrow \quad \swarrow{\scriptstyle \mathrm{proj}_1} & \\ & X & \end{array} .$$

If a finite covering map $\pi\colon E \to X$ can be embedded into the trivial complex line bundle over $X$, we can always assume – up to equivalence of covering maps – that $E$ is a subset of $X \times \mathbb{C}$ and that $\pi$ is given by projection onto $X$. We just have to identify $E$ with its image $h(E) \subset X \times \mathbb{C}$ under an embedding $h\colon E \to X \times \mathbb{C}$.

By its very definition, a polynomial covering map $\pi\colon E \to X$ is embedded into the trivial complex line bundle over $X$. If conversely, we are given an n–fold covering map $\pi\colon E \to X$ embedded into the trivial complex line bundle over $X$ such that $E \subset X \times \mathbb{C}$, then we can define a simple Weierstrass polynomial $P(x, z)$ over $X$ by associating to every $x \in X$ the polynomial of degree $n$,

$$P(x, z) = \prod_{(x, z_x) \in \pi^{-1}(x)} (z - z_x) ,$$

arising as the product of the linear polynomials $z - z_x$, where $(x, z_x)$ runs through the set of $n$ points in the fibre of $\pi$ over $x \in X$. Due to the local triviality of $\pi\colon E \to X$ it is clear, that the coefficient functions $a_1,\dots,a_n\colon X \to \mathbb{C}$ in $P(x, z)$ so defined will be continuous functions on $X$. Clearly, $\pi\colon E \to X$ is the polynomial covering map associated with $P(x, z)$.

Putting things together we have proved

Theorem 3.6. (The Embedding Criterion). A finite covering map $\pi\colon E \to X$ is equivalent to a polynomial covering map if and only if it admits an embedding into the trivial complex line bundle over $X$.

We finish this section with two applications of the embedding criterion. The first application provides us with the first example of a finite covering map, which is not polynomial.

Theorem 3.7. The standard double covering of the real projective n–space $\mathbb{RP}^n$ by the n–sphere $S^n$, $\pi: S^n \to \mathbb{RP}^n$, is not equivalent to a polynomial covering over $\mathbb{RP}^n$ when $n \geq 2$.

Proof. The proof is by contradiction. Suppose that $\pi: S^n \to \mathbb{RP}^n$ is equivalent to a polynomial covering map onto $\mathbb{RP}^n$. Then by Theorem 3.6 there is an embedding $h = (\pi, f): S^n \to \mathbb{RP}^n \times \mathbb{C}$ of $\pi$ into the trivial complex line bundle over $\mathbb{RP}^n$. The second factor of such an embedding provides a map $f: S^n \to \mathbb{C}$ for which $f(x) \neq f(-x)$ for any pair of antipodal points $x, -x \in S^n$ on the n–sphere. For $n \geq 2$ this contradicts the classical Borsuk–Ulam theorem, thereby proving Theorem 3.7.

For $n = 1$, we have $\mathbb{RP}^1 = S^1$, and $\pi: S^1 \to \mathbb{RP}^1 = S^1$ is just the nontrivial double covering of the circle. This covering was shown to be polynomial in Example 1.4.

The second application of the embedding criterion is to the orientation coverings of the nonorientable surfaces by the orientable surfaces.

Let $T_g$ denote the closed orientable surface of genus $g \geq 0$, i.e. a sphere with g handles, and let $U_g$ denote the closed nonorientable surface of genus $g \geq 1$, i.e. a sphere with g crosscaps.

It is well known that $T_g$ doubly covers $U_{g+1}$ for all $g \geq 0$. In fact, $U_{g+1}$ has $T_g$ as orientation covering space. Up to equivalence this double covering can be constructed as follows.

Consider $T_g$ as an embedded surface in euclidean 3–space $\mathbb{E}^3$, embedded such that the surface is symmetric with respect to reflection in the origin $0 \in \mathbb{E}^3$. See Figure 3. The antipodal map on $\mathbb{E}^3$ induces then an involution on $T_g$. The orbit space for this involution is exactly $U_{g+1}$, and the canonical projection of $T_g$ onto the orbit space $U_{g+1}$ is the covering map $p_g: T_g \to U_{g+1}$ in question. This is the model we use in the following.

For this covering we have

Theorem 3.8. The 2–fold covering map $p_g: T_g \to U_{g+1}$ for $g \geq 0$ is equivalent to a polynomial covering map if and only if g is odd.

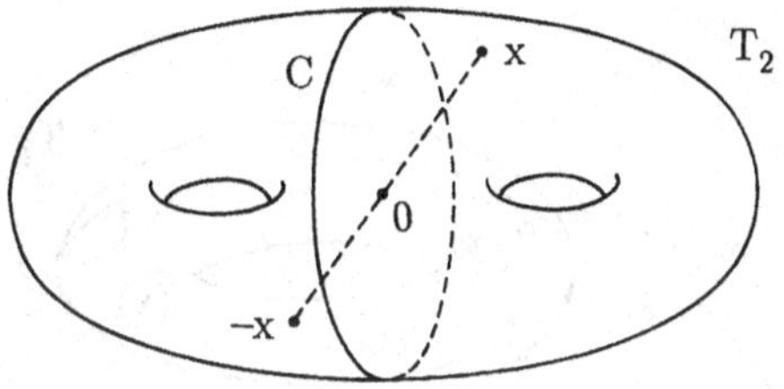


Figure 3

The proof of Theorem 3.8 proceeds as follows. By Theorem 3.6 we know that $p_g: T_g \to U_{g+1}$ is equivalent to a polynomial covering map if and only if there is an embedding $h = (p_g, f): T_g \to U_{g+1} \times \mathbb{C}$ of $p_g$ into the trivial complex line bundle over $U_{g+1}$. On the other hand such an embedding exists if and only if there exists a continuous map $f: T_g \to \mathbb{C}$, for which $f(x) \neq f(-x)$ for all $x \in T_g$, i.e. $f$ takes different values on pairs of antipodal points.

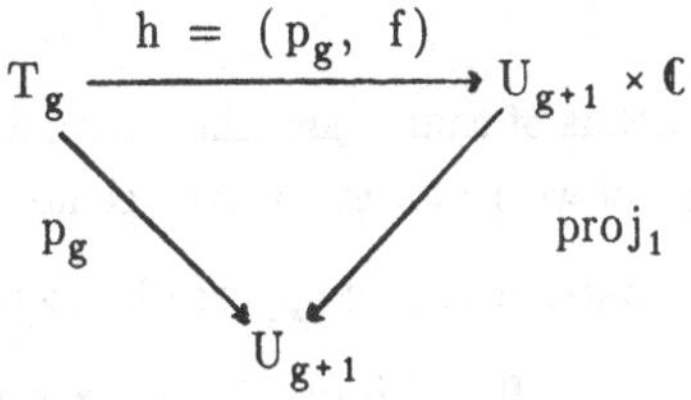


Theorem 3.8 is now a consequence of Lemma 3.9 and Lemma 3.10 below, of which the last lemma has independent interest.

Lemma 3.9. For $g$ odd, there exists a continuous map $f: T_g \to \mathbb{C}$, such that $f(x) \neq f(-x)$ for all $x \in T_g$.

Proof. Since $T_g$ has an odd number of holes, we can choose a plane $P$ through the origin $0 \in \mathbb{E}^3$, such that the line through $0 \in \mathbb{E}^3$ perpendicular to $P$ is disjoint from $T_g$. See Figure 4. It is clear that orthogonal projection of $T_g$ into $P$ provides a continuous map $f: T_g \to \mathbb{C}$ as required in Lemma 3.9, when we identify $P$ with $\mathbb{C}$.

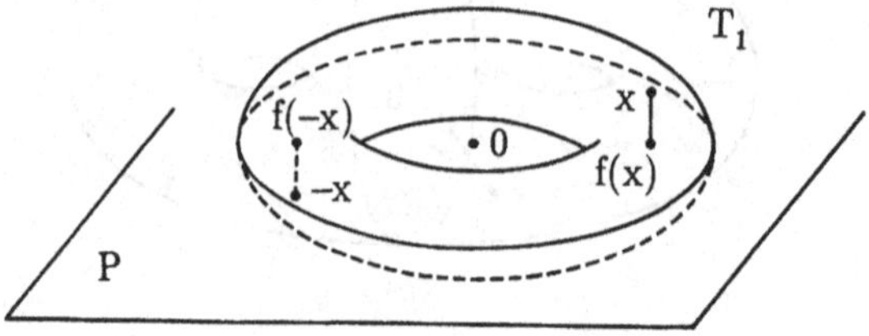


Figure 4

Lemma 3.10. (A Borsuk–Ulam theorem for orientable surfaces of even genus). Suppose that $g \geq 0$ is even. Then for every continuous map $f: T_g \to \mathbb{C}$ there exists a point $x \in T_g$ such that $f(x) = f(-x)$.
Equivalently: There does not exist a continuous map $f: T_g \to \mathbb{C}$ such that $f(x) \neq f(-x)$ for all $x \in T_g$ when $g \geq 0$ is even.

Proof. For $g = 0$, Lemma 3.10 is of course just the classical Borsuk–Ulam theorem in dimension 2. The following proof works for every even genus $g \geq 0$.

As usual the proof is by contradiction. Suppose therefore that there exists a continuous map $f: T_g \to \mathbb{C}$ such that $f(x) \neq f(-x)$ for all $x \in T_g$. We can then define a continuous map $h: T_g \to S^1$ by

$$h(x) = \frac{f(x) - f(-x)}{|f(x) - f(-x)|}$$

for $x \in T_g$. By construction $h(-x) = -h(x)$ for every $x \in T_g$, i.e. $h$ is an odd map.

Since $T_g$ has an even number of holes, there exists a simple, closed curve $C$ on $T_g$, which is invariant under the antipodal map, and which bounds a 2–chain in the homology of $T_g$. See Figure 3. The restriction of $h$ to $C$ defines an odd map $h|C: C \to S^1$ from the "circle" $C$ to $S^1$. It is then well known that $h|C$ has odd degree.

On the other hand, since the cycle represented by $C$ bounds in the homology of $T_g$, the image cycle $h(C)$ bounds in the homology of $S^1$, and therefore the degree of $h|C$ must be zero. Since this contradicts that $h|C$ has odd degree, Lemma 3.10 follows.

## 4. Polynomial covering maps and homomorphisms into braid groups.

There is a close connection between polynomial covering maps and homomorphisms into the Artin braid groups. This connection will be explored in the following three sections.

Recall from §2 that there is a canonical homeomorphism

$$\Phi\colon C_n(\mathbb{C}) \to B^n ,$$

which to the unordered n–tuple $[\alpha_1,\dots,\alpha_n]$ of n pairwise different complex numbers $\alpha_1,\dots,\alpha_n$ associates the ordered set of coefficients $(a_1,\dots,a_n) \in B^n = \mathbb{C}^n\backslash\Delta$ for the polynomial having $\alpha_1,\dots,\alpha_n$ as roots, i.e.

$$\Phi([\alpha_1,\dots,\alpha_n]) = (a_1,\dots,a_n) ,$$

when
$$z^n + \sum_{i=1}^{n} a_i\, z^{n-i} = \prod_{i=1}^{n} (z - \alpha_i) .$$

As usual, $p_n\colon F_n(\mathbb{C}) \to C_n(\mathbb{C})$ denotes the principal $\Sigma_n$ – bundle of configuration spaces.

Choose base points in the above spaces as follows:

$$\bar{c}_0 = (1, 2,\dots,n) \in F_n(\mathbb{C})$$
$$c_0 = p_n(\bar{c}_0) \in C_n(\mathbb{C})$$
$$b_0 = \Phi(c_0) \in B^n .$$

Then we know by Theorem I.3.1, that the Artin braid group $B(n)$ can be canonically identified with the fundamental group $\pi_1(C_n(\mathbb{C}), c_0)$, which on the other hand is isomorphic to the fundamental group $\pi_1(B^n, b_0)$ under the isomorphism induced by $\Phi$. Altogether we have the identifications

$$B(n) = \pi_1(C_n(\mathbb{C}), c_0) = \pi_1(B^n, b_0) .$$

Now let $\pi_a\colon E_a \to X$ be an n–fold polynomial covering map with coefficient map $a\colon X \to B^n$. Then we can define a unique map $\alpha\colon X \to C_n(\mathbb{C})$ such that the following diagram is commutative,

$$\begin{array}{ccc} & & C_n(\mathbb{C}) \\ & \overset{\alpha}{\nearrow} & \\ X & & \downarrow \Phi \\ & \underset{a}{\searrow} & \\ & & B^n \end{array} .$$

For obvious reasons, $\alpha\colon X \to C_n(\mathbb{C})$ is called <u>the root map</u> for the polynomial covering map $\pi_a\colon E_a \to X$, or for the underlying simple Weierstrass polynomial $P_a(x, z)$.

Clearly the root map $\alpha$ and the coefficient map $a$ for a polynomial covering map mutually determine each other via the homeomorphism $\Phi$.

Let $x_0 \in X$ be the base point in $X$. Then the root map $\alpha\colon X \to C_n(\mathbb{C})$ induces a homomorphism

$$\alpha_*\colon \pi_1(X, x_0) \to \pi_1(C_n(\mathbb{C}), \alpha(x_0))$$

between fundamental groups. Since $C_n(\mathbb{C})$ is pathwise connected and $c_0 \in C_n(\mathbb{C})$ is a nondegenerate base point in $C_n(\mathbb{C})$, this homomorphism determines a unique conjugacy class of homomorphisms of $\pi_1(X)$ into the Artin braid group $B(n) = \pi_1(C_n(\mathbb{C}), c_0)$. Any one of the homomorphisms in this conjugacy class – or the conjugacy class itself – will also be denoted by

$$\alpha_*\colon \pi_1(X) \to B(n) .$$

Let $\mathrm{Hom}(\pi_1(X), B(n))^{\mathrm{conj}}$ denote the set of conjugacy classes of homomorphisms of $\pi_1(X)$ into $B(n)$.

Assume from now on, that $X$ has the homotopy type of a CW–complex.

For an arbitrary topological space $Y$, we denote by $[X, Y]$ the set of free homotopy classes of maps of $X$ into $Y$.

According to Theorem I.3.2, the configuration space $C_n(\mathbb{C})$ is an Eilenberg–MacLane space of type $(B(n), 1)$. By a well known theorem from algebraic topology (see e.g. [12], Theorem 11, p. 428), there is therefore a bijective correspondence

$$[X, C_n(\mathbb{C})] \to \mathrm{Hom}(\pi_1(X), B(n))^{\mathrm{conj}} ,$$

defined by mapping the free homotopy class of a root map $\alpha\colon X \to C_n(\mathbb{C})$ into the conjugacy class of the induced homomorphism $\alpha_*\colon \pi_1(X) \to B(n)$.

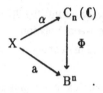

The homeomorphism $\Phi\colon C_n(\mathbb{C}) \to B^n$ induces a bijective correspondence

$$\Phi_*\colon [X, C_n(\mathbb{C})] \to [X, B^n] ,$$

which maps the free homotopy class of a root map $\alpha\colon X \to C_n(\mathbb{C})$ into the free homotopy class of the corresponding coefficient map $a\colon X \to B^n$ .

Let $\mathscr{PC}_n(X)$ denote the set of equivalence classes of n–fold polynomial covering maps onto $X$ . By Corollary 3.3 there is then a well defined surjective map

$$[X, B^n] \to \mathscr{PC}_n(X) \ ,$$

which to the free homotopy class of a map $a\colon X \to B^n$ associates the equivalence class of the n–fold polynomial covering map $\pi_a\colon E_a \to X$ with coefficient map $a$ .

The above remarks are summarized in the diagram below. Using the diagram we can freely switch between presenting an equivalence class of n–fold polynomial covering maps onto $X$ by a free homotopy class of maps of $X$ into $B^n$ , or a free homotopy class of maps of $X$ into $C_n(\mathbb{C})$ , or a conjugacy class of homomorphisms of $\pi_1(X)$ into $B(n)$ .

| | | | |
|---|---|---|---|
| Conjugacy class of $\alpha_*\colon \pi_1(X) \to B(n)$ | $\mathrm{Hom}(\pi_1(X), B(n))^{\mathrm{conj}}$ | | |
| | $\uparrow$ bijective correspondence | | |
| Free homotopy class of $\alpha\colon X \to C_n(\mathbb{C})$ | $[X, C_n(\mathbb{C})]$ | | |
| | $\uparrow$ bijective correspondence | | |
| Free homotopy class of $a\colon X \to B^n$ | $[X, B^n]$ $\xrightarrow{\text{surjective map}}$ | $\mathscr{PC}_n(X)$ | Equivalence class of $\pi_a\colon E_a \to X$ |

The algebraic description of polynomial covering maps in terms of conjugacy classes of homomorphisms into the Artin braid groups has the following immediate application.

<u>Theorem 4.1.</u> Suppose that $\pi_1(X)$ is a torsion group, i.e. all nontrivial elements in $\pi_1(X)$ have finite order. Then every polynomial covering map onto $X$ is trivial.

Proof. By Corollary I.3.3 we know that the Artin braid group $B(n)$ contains no nontrivial elements of finite order. If $\pi_1(X)$ is a torsion group, there is therefore only one homomorphism of $\pi_1(X)$ into $B(n)$, namely the trivial homomorphism. Therefore $\mathscr{PC}_n(X)$ contains only one element, namely the equivalence class of the trivial n–fold covering map onto $X$. This proves Theorem 4.1.

As a particular case we can take $X$ to be the real projective n–space $\mathbb{R}P^n$. For $n \geq 2$, the fundamental group of $\mathbb{R}P^n$ is the cyclic group of order 2. Therefore by Theorem 4.1, $\mathbb{R}P^n$ does not admit any nontrivial polynomial covering spaces. In particular, it follows that the standard double covering $\pi: S^n \to \mathbb{R}P^n$ of $\mathbb{R}P^n$ by the n–sphere $S^n$ – which is a nontrivial covering map – is not equivalent to a polynomial covering map for $n \geq 2$. Thereby we get an alternative proof of Theorem 3.7.

The surjective map $[X, B^n] \to \mathscr{PC}_n(X)$ is not in general injective. In fact, Jesper Michael Møller has proved in [39], see also Theorem IV.1.7, that if and only if $H^1(X; \mathbb{Z}) = 0$ does $X$ not allow a homotopically nontrivial map $a: X \to B^n$ with corresponding trivial polynomial covering map $\pi_a: E_a \to X$. On the other hand spaces $X$ with $H^1(X; \mathbb{Z}) = 0$ are apt to have few coverings – in particular few polynomial coverings. Here is a concrete example of a space $X$ and a homotopically nontrivial map $a: X \to B^n$ with corresponding trivial polynomial covering map $\pi_a: E_a \to X$.

Example 4.2. Let $S^1$ be the unit circle, considered as the set of complex numbers $x \in \mathbb{C}$ of modulus $|x| = 1$.

For the simple Weierstrass polynomial

$$P(x, z) = z^2 - x^2 \ , \ x \in S^1 \ , \ z \in \mathbb{C} \ ,$$

over $S^1$, the coefficient map $a = (a_1, a_2): S^1 \to B^2 = \mathbb{C}^2 \setminus \Delta$ and the root map $\alpha: S^1 \to C_2(\mathbb{C})$ are given respectively by $a(x) = (0, -x^2)$ and $\alpha(x) = [x, -x]$ for $x \in S^1$.

The polynomial covering map $\pi: E \to S^1$ associated with $P(x, z)$ is trivial. An explicit trivialization,

$$\begin{array}{ccc} S^1 \times \{-1, 1\} & \xrightarrow{h} & E \\ & {}_{\mathrm{proj}_1}\searrow \quad \swarrow_{\pi} & \\ & S^1 & \end{array} \ ,$$

can be defined by $h(x, \pm 1) = (x, \pm x)$ .

We shall now prove that the coefficient map $a\colon S^1 \to B^2$ , or equivalently, the root map $\alpha\colon S^1 \to C_2(\mathbb{C})$ , is not freely homotopic to a constant map.

First observe, that since $F_2(\mathbb{C}) = \{(z_1, z_2) \in \mathbb{C}^2 \mid z_1 \neq z_2\}$ , there is a homeomorphism $F_2(\mathbb{C}) \to \mathbb{C} \times (\mathbb{C}\backslash\{0\})$ , which maps $(z_1, z_2) \in F_2(\mathbb{C})$ into $(\zeta_1, \zeta_2) = (z_1 + z_2, z_1 - z_2)$ . The root map $\alpha\colon S^1 \to C_2(\mathbb{C})$ admits the lifting $\alpha'\colon S^1 \to F_2(\mathbb{C})$ , $\alpha'(x) = (x, -x)$ , over the covering map $p_2\colon F_2(\mathbb{C}) \to C_2(\mathbb{C})$ . Under the above homeomorphism, $\alpha'\colon S^1 \to F_2(\mathbb{C})$ obviously corresponds to the map $\tilde{\alpha}'\colon S^1 \to \mathbb{C} \times (\mathbb{C}\backslash\{0\})$ defined by $\tilde{\alpha}'(x) = (0, 2x)$ .

There is an obvious homotopy equivalence $\mathbb{C} \times (\mathbb{C}\backslash\{0\}) \simeq S^1$ , defined by radial projection in the second argument. Since $S^1$ is a simple space (it is a topological group), we can work with based homotopy classes instead of free homotopy classes of maps into $S^1$ .

Using the above remarks it follows that there is a sequence of isomorphisms

$$\pi_1(F_2(\mathbb{C})) \cong \pi_1(\mathbb{C} \times (\mathbb{C}\backslash\{0\})) \cong \pi_1(S^1) \cong \mathbb{Z}$$

to the integers $\mathbb{Z}$ , and that $\alpha'\colon S^1 \to F_2(\mathbb{C})$ through these isomorphisms corresponds to $1 \in \mathbb{Z}$ . Hence $\alpha'\colon S^1 \to F_2(\mathbb{C})$ is not freely homotopic to a constant map. Since $p_2\colon F_2(\mathbb{C}) \to C_2(\mathbb{C})$ is a covering map, this implies that the root map $\alpha\colon S^1 \to C_2(\mathbb{C})$ is not freely homotopic to a constant map. This finishes Example 4.2.

Finally in this section a few words about classification of the polynomial covering maps.

According to Theorem 3.1, the surjective map $[X, B^n] \to \mathcal{PC}_n(X)$ can also be defined as the map, which to the free homotopy class of $a\colon X \to B^n$ , associates the equivalence class of the pull–back of the canonical n–fold polynomial covering map $\pi^n\colon E^n \to B^n$ along $a$ . Since the map described is not in general injective, the canonical n–fold polynomial covering map $\pi^n\colon E^n \to B^n$ is not a classifying object for the n–fold polynomial covering maps. But more than that is true, since Jesper Michael Møller [39] has shown that there does not exist any n–fold polynomial covering map

$w^n: W^n \to Z^n$ for $n \geq 2$ such that the map $[X, Z^n] \to \mathscr{PC}_n(X)$ defined by pull–back of $w^n$ is bijective for all CW–complexes $X$. [As observed by Anatoly Libgober in Math. Reviews No. 83b55013, this also follows from the failure of the Mayer – Vietoris property for the functor $\mathscr{PC}_n(X)$, which therefore is not representable in the sense of Brown: Ann. of Math. (2) 75 (1962), 467–484.] Hence in complete generality, there is no hope for a geometric classification in terms of classifying spaces for the polynomial covering maps.

We shall return to describe an algebraic classification of the polynomial covering maps in §6.

## 5. Characteristic homomorphisms for finite covering maps.

As usual $X$ denotes a connected and locally pathwise connected topological space. In particular, $X$ is then pathwise connected. Choose a base point $x_0 \in X$ in $X$.

Let $\pi: E \to X$ be an arbitrary n–fold covering map onto $X$, and choose an ordering of the points in the fibre over the base point $x_0 \in X$. Say that $\pi^{-1}(x_0) = \{z_1,\ldots,z_n\}$.

Path lifting in $\pi: E \to X$ defines a homomorphism

$$\chi(\pi): \pi_1(X) \to \Sigma_n$$

of the fundamental group $\pi_1(X)$ of $X$ at $x_0 \in X$ into the symmetric group on $n$ elements $\Sigma_n$.

The definition goes as follows. Given a loop $\ell: [0,1] \to X$ in $X$ based at $x_0 \in X$, i.e. $\ell(0) = \ell(1) = x_0$. Then there is a unique lifting of $\ell$ to a path $\ell_i: [0,1] \to E$ in $E$ with $\ell_i(0) = z_i$ for each $i = 1,\ldots,n$. Since $\ell_i(1)$ is again a point in the fibre $\pi^{-1}(x_0)$, these liftings define a permutation $\tau \in \Sigma_n$, such that $\ell_i(1) = z_{\tau(i)}$ for each $i = 1,..,n$. It is easy to prove that $\tau$ only depends on the homotopy class of the loop $\ell$, and that the assignment of $\tau$ to the homotopy class of $\ell$, actually defines a homomorphism $\chi(\pi)$ as asserted.

If we change base point $x_0 \in X$ in $X$, or change the ordering of the points in the fibre $\pi^{-1}(x_0)$ over the base point, then this will change $\chi(\pi)$ by a conjugation in $\Sigma_n$, i.e. change $\chi(\pi)$ to $\tau_0\, \chi(\pi)\, \tau_0^{-1}$ for a fixed permutation $\tau_0 \in \Sigma_n$.

We call $\chi(\pi): \pi_1(X) \to \Sigma_n$ the characteristic homomorphism for the n–fold covering map $\pi: E \to X$. As we have seen, the characteristic homomorphism is only well defined up to a conjugation in $\Sigma_n$.

The name characteristic homomorphism is appropriate, since we have

Theorem 5.1. Two n–fold covering maps $\pi_i: E_i \to X$, $i = 1,2,$ onto $X$ are equivalent if and only if their characteristic homomorphisms $\chi(\pi_i): \pi_1(X) \to \Sigma_n$, $i = 1,2,$ are conjugate homomorphisms.

Proof. First suppose that $\pi_i: E_i \to X$, $i = 1,2,$ are equivalent covering maps,

$$\begin{array}{ccc} E_1 & \xrightarrow{h} & E_2 \\ & {\scriptstyle\pi_1}\searrow \quad \swarrow{\scriptstyle\pi_2} & \\ & X & \end{array} .$$

If $\pi_1^{-1}(x_0) = \{z_1^1,...,z_n^1\}$ is the ordering of the points in the fibre over $x_0$ in $\pi_1$, then take $\pi_2^{-1}(x_0) = \{z_1^2,...,z_n^2\} = \{h(z_1^1),...,h(z_n^1)\}$ to be the ordering of the points in the fibre over $x_0$ in $\pi_2$.

Since $h$ maps paths in $E_1$, lifting loops in $X$, to paths in $E_2$, lifting these same loops in $X$, it follows easily that $\chi(\pi_1) = \chi(\pi_2)$, and hence that the characteristic homomorphisms of $\pi_1$ and $\pi_2$ are conjugate homomorphisms.

Suppose next that $\chi(\pi_i): \pi_1(X) \to \Sigma_n$, $i = 1,2,$ are conjugate homomorphisms. After a possible reordering of the points in $\pi_2^{-1}(x_0) = \{z_1^2,...,z_n^2\}$ we can assume that $\chi(\pi_1) = \chi(\pi_2)$.

We shall define an equivalence

$$\begin{array}{ccc} E_1 & \xrightarrow{h} & E_2 \\ & {\scriptstyle\pi_1}\searrow \quad \swarrow{\scriptstyle\pi_2} & \\ & X & \end{array} .$$

Start out by setting $h(z_i^1) = z_i^2$ for $i = 1,...,n$.

For every point $e_1 \in E_1$, there exists a path $c_1: [0,1] \to E_1$ such that $c_1(0) = z_i^1$ for some $i = 1,...,n$ and $c_1(1) = e_1$. This is so, since $X$ is pathwise connected and $\pi_1$ has the path lifting property. By the path lifting property for $\pi_2$, there is then a unique path $c_2: [0,1] \to E_2$ such that $\pi_1 \circ c_1 = \pi_2 \circ c_2$ and $c_2(0) = z_i^2$. Now define $h(e_1) = c_2(1)$.

This map is well defined. To prove this, suppose $d_1: [0,1] \to E_1$ is another path in $E_1$ with $d_1(0) = z_j^1$ for some $j = 1,...,n$ and $d_1(1) = e_1$. Together, the projections of the paths $c_1$ and $d_1$ under $\pi_1$ define a loop $\ell: [0,1] \to X$ in $X$, $\ell(0) = \ell(1) = x_0$, such that $z_j^1 = \chi(\pi_1)(\ell)(z_i^1)$. The unique lift $\ell': [0,1] \to E_1$ of $\ell$ to $E_1$ with $\ell'(0) = z_i^1$ satisfies in fact $\ell'(1) = z_j^1$, and furthermore, $c_1$ is homotopic to the product path $\ell' \cdot d_1$; we write $c_1 \sim \ell' \cdot d_1$. See Figure 5. But then $c_2 \sim \ell'' \cdot d_2$, where $\ell'': [0,1] \to E_2$ is the unique lift of the loop $\ell$ to $E_2$ with $\ell''(0) = z_i^2$ and $\ell''(1) = z_j^2$, since $\chi(\pi_1) = \chi(\pi_2)$, and $d_2$ is defined in analogy with $c_2$. We conclude that $d_2(0) = z_j^2$ and in particular that $d_2(1) = c_2(1)$, which proves that the map

$h: E_1 \to E_2$ is well defined.

Since X is locally pathwise connected, it is easy to prove that $h: E_1 \to E_2$ is continuous. By construction, h commutes with projections onto X . The inverse procedure defines an inverse map $h^{-1}: E^2 \to E^1$ to h . Altogether it follows, that $h: E_1 \to E_2$ is an equivalence of covering maps.

This proves Theorem 5.1.

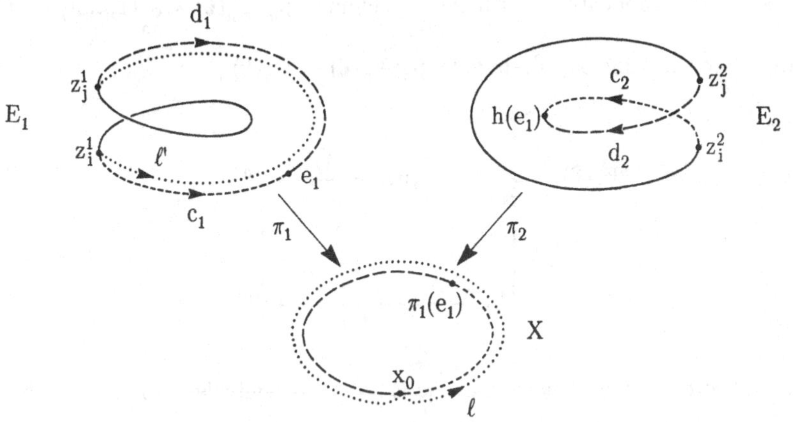


Figure 5

Suppose now that $\pi: E \to X$ is an n–fold polynomial covering map with root map $\alpha: X \to C_n(\mathbb{C})$ . In this case we can give an alternative description of the characteristic homomorphism $\chi(\pi): \pi_1(X) \to \Sigma_n$ of $\pi$ , very useful, it turns out.

As explained in §4, the root map $\alpha: X \to C_n(\mathbb{C})$ for $\pi$ induces a well defined conjugacy class of homomorphisms $\alpha_*: \pi_1(X) \to B(n)$ . Compose a homomorphism $\alpha_*$ in this conjugacy class with the permutation homomorphism $\tau_n: B(n) \to \Sigma_n$ from the braid group sequence

$$1 \longrightarrow H(n) \xrightarrow{\rho_n} B(n) \xrightarrow{\tau_n} \Sigma \longrightarrow 1$$

defined in Chapter I, §3. Thereby we get a well defined conjugacy class of homomorphisms $\tau_n \circ \alpha_*: \pi_1(X) \to \Sigma_n$ . As the following theorem shows, this is exactly the characteristic homomorphism of $\pi$ .

Theorem 5.2. Let $\pi\colon E \to X$ be an n–fold polynomial covering map with root map $\alpha\colon X \to C_n(\mathbb{C})$ . Then the characteristic homomorphism $\chi(\pi)$ of $\pi$ is given by $\chi(\pi) = \tau_n \circ \alpha_*$ .

The formula $\chi(\pi) = \tau_n \circ \alpha_*$ shall be understood as an equality of conjugacy classes of homomorphisms of $\pi_1(X)$ into $\Sigma_n$ .

Proof. Recall from §2 that the canonical n–fold polynomial covering map $\pi^n\colon E^n \to B^n$ can be identified with the $\Sigma_n$–bundle $\bar{p}_n\colon F_n(\mathbb{C}) \times_{\Sigma_n} \{1,\dots,n\} \to C_n(\mathbb{C})$ associated with the principal $\Sigma_n$–bundle $p_n\colon F_n(\mathbb{C}) \to C_n(\mathbb{C})$ ,

$$\begin{array}{ccc} F_n(\mathbb{C}) \times_{\Sigma_n} \{1,\dots,n\} & \xrightarrow{\ \Phi\ } & E^n \\ {\scriptstyle \bar{p}_n}\downarrow & & \downarrow{\scriptstyle \pi^n} \\ C_n(\mathbb{C}) & \xrightarrow[\ \Phi\ ]{} & B^n \end{array}$$

From Theorem 3.1 we know that $\pi\colon E \to X$ is equivalent to the pull–back of $\pi^n\colon E^n \to B^n$ along the coefficient map $a = \Phi \circ \alpha\colon X \to B^n$ for $\pi$ . By the above diagram it follows that $\pi\colon E \to X$ is equivalent to the pull–back of $\bar{p}_n\colon F_n(\mathbb{C}) \times_{\Sigma_n} \{1,\dots,n\} \to C_n(\mathbb{C})$ along the root map $\alpha\colon X \to C_n(\mathbb{C})$ for $\pi$ . We have in other words a commutative diagram

$$\begin{array}{ccc} E & \xrightarrow{\ \alpha'\ } & F_n(\mathbb{C}) \times_{\Sigma_n} \{1,\dots,n\} \\ {\scriptstyle \pi}\downarrow & & \downarrow{\scriptstyle \bar{p}_n} \\ X & \xrightarrow[\ \alpha\ ]{} & C_n(\mathbb{C}) \end{array} ,$$

in which $\alpha'$ is bijective on the fibres of $\pi$ .

The characteristic homomorphism for $\bar{p}_n$ is exactly the boundary homomorphism in the homotopy sequence for the principal bundle $p_n$ , i.e. $\chi(\bar{p}_n) = \tau_n\colon B(n) \to \Sigma_n$ .

This follows since both homomorphisms are defined by path lifting, and liftings of loops in $C_n(\mathbb{C})$ to paths in $F_n(\mathbb{C}) \times_{\Sigma_n} \{1,...,n\}$ over $\overline{p}_n$ are induced from liftings to paths in $F_n(\mathbb{C})$ over $p_n$.

Since path liftings in $\pi$ are pushed over to path liftings in $\overline{p}_n$ by the pair of maps $(\alpha, \alpha')$, it follows now immediately, that $\chi(\pi) = \tau_n \circ \alpha_*$.

This proves Theorem 5.2.

## 6. An algebraic classification of the polynomial covering maps.

Throughout this section we assume that $X$ is a connected, locally pathwise connected and semilocally simply connected topological space. These conditions are imposed on $X$ in order to ensure that $X$ has a universal covering space. We shall further assume that $X$ has the homotopy type of a CW–complex, such that the equivalence classes of n–fold polynomial covering maps onto $X$ are described by conjugacy classes of homomorphisms of $\pi_1(X)$ into the Artin braid group $B(n)$ as in §4.

In the main results in this section we shall provide necessary and sufficient conditions, for a homomorphism of $\pi_1(X)$ into $\Sigma_n$ to represent the equivalence class of an n–fold polynomial covering map onto $X$, and for two homomorphisms of $\pi_1(X)$ into $B(n)$ to define equivalent polynomial covering maps onto $X$.

Most of our investigations are related to the spaces and maps, which will be defined subsequently, in the following commutative diagram,

$$\begin{array}{ccc} \mathrm{Hom}(\pi_1(X), H(n)) & & \\ \rho_n \circ - \downarrow & & \\ \mathrm{Hom}(\pi_1(X), B(n)) & \longrightarrow & \mathscr{PC}_n(X) \\ \tau_n \circ - \downarrow & & \downarrow \\ \mathrm{Hom}(\pi_1(X), \Sigma_n) & \longrightarrow & \mathscr{C}_n(X) \end{array} .$$

$\mathscr{C}_n(X)$ and $\mathscr{PC}_n(X)$ denote respectively the set of equivalence classes of n–fold covering maps and n–fold polynomial covering maps onto $X$. There is an inclusion map $\mathscr{PC}_n(X) \hookrightarrow \mathscr{C}_n(X)$.

For any pair of groups $G$ and $H$, we denote by $\mathrm{Hom}(G, H)$ the set of homomorphisms of $G$ into $H$. The maps $\rho_n \circ -$ and $\tau_n \circ -$ in the diagram are induced by composition of homomorphisms from the homomorphisms in the braid group sequence

$$1 \to H(n) \xrightarrow{\rho_n} B(n) \xrightarrow{\tau_n} \Sigma_n \to 1 .$$

Finally, we explain now the horizontal maps in the diagram. The map

$$\mathrm{Hom}(\pi_1(X), \Sigma_n) \to \mathscr{C}_n(X)$$

is the surjective map, which to the homomorphism $\varphi\colon \pi_1(X) \to \Sigma_n$ associates the equivalence class of the n–fold covering map $\pi\colon E \to X$ having $\varphi$ as a characteristic homomorphism. The covering map $\pi\colon E \to X$ can be constructed as an associated bundle to the universal covering space map $\tilde{\pi}: \tilde{X} \to X$ viewed as a principal $\pi_1(X)$ – bundle. From Theorem 5.1 we know, that two homomorphisms $\varphi, \psi \in \mathrm{Hom}(\pi_1(X), \Sigma_n)$ determine the same equivalence class of n–fold covering maps onto X if and only if they are conjugate homomorphisms. In particular we note that the constant homomorphism of $\pi_1(X)$ into the identity permutation in $\Sigma_n$ is the unique representative for the equivalence class of the trivial n–fold covering map onto X .

The map

$$\mathrm{Hom}(\pi_1(X), B(n)) \to \mathscr{PC}_n(X)$$

is the surjective map, which to the homomorphism $\varphi: \pi_1(X) \to B(n)$ associates the equivalence class of the n–fold polynomial covering map $\pi: E \to X$ having $\chi(\pi) = \tau_n \circ \varphi$ as a characteristic homomorphism. That such a map exists follows by the constructions in §4 combined with Theorem 5.2.

This finishes the definitions of the spaces and maps in the above diagram.

First we make use of the diagram to prove

Theorem 6.1. The equivalence class of an n–fold covering map $\pi: E \to X$ represented by the homomorphism $\varphi: \pi_1(X) \to \Sigma_n$ contains a polynomial covering map if and only if there exists a homomorphism $\varphi': \pi_1(X) \to B(n)$ such that $\varphi = \tau_n \circ \varphi'$ .

In different words: An n–fold covering map $\pi: E \to X$ is equivalent to a polynomial covering map if and only if the characteristic homomorphism $\chi(\pi): \pi_1(X) \to \Sigma_n$ of $\pi$ lifts over $\tau_n: B(n) \to \Sigma_n$ , i.e. $\chi(\pi) = \tau_n \circ \varphi'$ for a homomorphism $\varphi': \pi_1(X) \to B(n)$ .

Proof. An n–fold polynomial covering map onto X is represented by a homomorphism $\psi: \pi_1(X) \to B(n)$ . The homomorphisms $\varphi$ , $\tau_n \circ \psi \in \mathrm{Hom}(\pi_1(X), \Sigma_n)$ represent equivalent covering maps onto X if and only if they are conjugate homomorphisms, i.e. if and only if there exists an $s \in \Sigma_n$ such that $\varphi = s \cdot (\tau_n \circ \psi) \cdot s^{-1}$ . Since $\tau_n$ is an epimorphism we can choose $b \in B(n)$ such that $s = \tau_n(b)$ . Then it follows easily, that $\varphi$ and $\tau_n \circ \psi$ represent equivalent covering maps if and only if there exists a homomorphism $\varphi': \pi_1(X) \to B(n)$ , conjugate to $\psi: \pi_1(X) \to B(n)$ , such that $\varphi = \tau_n \circ \varphi'$ . We just have to take $\varphi' = b \cdot \psi \cdot b^{-1}$ . Theorem 6.1 follows now immediately using the main diagram.

Corollary 6.2. In order that the space X has the property that all n–fold covering maps onto X are equivalent to polynomial covering maps, it is necessary and sufficient that the map

$$\tau_n \circ - : \mathrm{Hom}(\pi_1(X), B(n)) \to \mathrm{Hom}(\pi_1(X), \Sigma_n)$$

is surjective.

If $\pi_1(X)$ is a free group, the condition in Corollary 6.2 is obviously satisfied. This is the case, since given a homomorphism $\varphi: \pi_1(X) \to \Sigma_n$, we can just lift the images in $\Sigma_n$ of a set of generators for $\pi_1(X)$ and then extend to a homomorphism $\varphi': \pi_1(x) \to B(n)$. We can lift the images of the generators, since $\tau_n: B(n) \to \Sigma_n$ is an epimorphism. Thereby we get the following

Theorem 6.3. Suppose that $\pi_1(X)$ is a free group. Then every finite covering map onto X is equivalent to a polynomial covering map.

Specific examples of spaces X with free fundamental groups are provided e.g. by punctured surfaces.

Next we provide an algebraic criterion for an n–fold polynomial covering map to represent the trivial n–fold covering map.

Theorem 6.4. The n–fold polynomial covering map $\pi: E \to X$ defined by a homomorphism $\varphi': \pi_1(X) \to B(n)$ is equivalent to the trivial n–fold covering map onto X if and only if there exists a homomorphism $\varphi'': \pi_1(X) \to H(n)$ such that $\varphi' = \rho_n \circ \varphi''$.

Proof. The homomorphism $\varphi': \pi_1(X) \to B(n)$ represents the trivial n–fold covering map onto X if and only if $\varphi = \tau_n \circ \varphi'$ is conjugate to, and therefore equal to, the constant homomorphism of $\pi_1(X)$ into the identity permutation in $\Sigma_n$. By exactness of the braid group sequence, it follows easily that this is the case if and only if there exists a homomorphism $\varphi'': \pi_1(X) \to H(n)$ such that $\varphi' = \rho_n \circ \varphi''$. This proves Theorem 6.4.

We now turn to the problem when two homomorphisms $\varphi, \psi \in \mathrm{Hom}(\pi_1(X), B(n))$ determine the same equivalence class of n–fold polynomial covering maps onto X.

By Theorem 5.1, we know that two homomorphisms $\varphi, \psi \in \mathrm{Hom}(\pi_1(X), B(n))$ determine the same equivalence class of n–fold polynomial covering maps onto X if and only if $\tau_n \circ \varphi, \tau_n \circ \psi \in \mathrm{Hom}(\pi_1(X), \Sigma_n)$ are conjugate homomorphisms. Since $\tau_n$ is an epimorphism, this condition is satisfied if and only if there exists an element $b \in B(n)$ such that $\tau_n \circ \psi = \tau_n \circ \varphi^b$. Here $\varphi^b \in \mathrm{Hom}(\pi_1(X), B(n))$ denotes the conjugation of $\varphi$ by b, i.e., for every $\gamma \in \pi_1(X)$ we have $\varphi^b(\gamma) = b \cdot \varphi(\gamma) \cdot b^{-1}$. If $\tau_n \circ \psi = \tau_n \circ \varphi^b$, then exactness of the braid group sequence shows that there exists a unique map (not necessarily a homomorphism) $\theta: \pi_1(X) \to H(n)$ such that $(\rho_n \circ \theta) \cdot \psi = \varphi^b$. The multiplication on the left in this equation is pointwise multiplication of maps of $\pi_1(X)$ into $B(n)$.

Since $(\rho_n \circ \theta) \cdot \psi$ has to be a homomorphism of $\pi_1(X)$ into $B(n)$, the map $\theta$ has to satisfy a certain condition. To describe this condition, we note that the homomorphism $\psi: \pi_1(X) \to B(n)$ induces an action of $\pi_1(X)$ on $H(n)$ as follows. For $\gamma \in \pi_1(X)$ and $c \in H(n)$ we let $c^{\psi(\gamma)} \in H(n)$ denote the unique element, which exists by exactness of the braid group sequence, such that

$$\rho_n(c^{\psi(\gamma)}) = \psi(\gamma) \cdot \rho_n(c) \cdot \psi(\gamma)^{-1} .$$

The condition, the map $\theta: \pi_1(X) \to H(n)$ has to satisfy in order that $(\rho_n \circ \theta) \cdot \psi$ is a homomorphism of $\pi_1(X)$ into $B(n)$, is then easily seen – after a small computation – to be expressed in the following

Definition 6.5. Let $\psi: \pi_1(X) \to B(n)$ be a homomorphism. A map $\theta: \pi_1(X) \to H(n)$ is then called a $\psi$ – crossed homomorphism if

$$\theta(\gamma_1 \cdot \gamma_2) = \theta(\gamma_1) \cdot \theta(\gamma_2)^{\psi(\gamma_1)}$$

for all $\gamma_1, \gamma_2 \in \pi_1(X)$.

Using the notation above, the arguments preceeding Definition 6.5 proves the following

Theorem 6.6. Two homomorphisms $\varphi, \psi \in \mathrm{Hom}(\pi_1(X), B(n))$ determine the same equivalence class of n–fold polynomial covering maps onto X if and only if there exists an element $b \in B(n)$ and a $\psi$ – crossed homomorphism $\theta: \pi_1(X) \to H(n)$ such that $(\rho_n \circ \theta) \cdot \psi = \varphi^b$

Theorem 6.1 and Theorem 6.6 taken together provide a satisfactory algebraic description of the equivalence classes of polynomial covering maps onto $X$. As was mentioned at the end of §4, there is no hope for a geometric classification in terms of classifying spaces for the polynomial covering maps.

We finish this section with a further application of Theorem 6.1.

Example 6.7. Let $p_g\colon T_g \to U_{g+1}$ denote the double covering of the nonorientable closed surface $U_{g+1}$ of genus $g+1$ by the orientable closed surface $T_g$ of genus $g \geq 0$ described in §3.

For the braid group on 2 strings $B(2)$ and the permutation group on 2 elements $\Sigma_2$, we have the isomorphisms

$$B(2) \cong \mathbb{Z} \quad \text{and} \quad \Sigma_2 \cong \mathbb{Z}/2$$

to the integers $\mathbb{Z}$ and the cyclic group $\mathbb{Z}/2$ of order 2. Under these identifications, the permutation homomorphism $\tau_2\colon B(2) \to \Sigma_2$ corresponds to reduction modulo 2.

It is well known [9] that the fundamental group of $U_{g+1}$ admits the presentation

$$\pi_1(U_{g+1}) = \langle c_1,\ldots,c_{g+1} \mid c_1^2\cdot\ldots\cdot c_{g+1}^2 = 1\rangle .$$

Since $p_g\colon T_g \to U_{g+1}$ is nontrivial, and since all generators $c_i \in \pi_1(U_{g+1})$ appear in a symmetric fashion, it is clear that the characteristic homomorphism of $p_g$,

$$\chi(p_g)\colon \pi_1(U_{g+1}) \to \Sigma_2 ,$$

is given by $\chi(p_g)(c_i) = \tau$, where $\tau$ is the nontrivial transposition generating $\Sigma_2$.

A lift $\varphi'\colon \pi_1(U_{g+1}) \to B(2)$ of $\chi(p_g)$ over $\tau_2$,

$$\begin{array}{ccc} & \overset{\varphi'}{\nearrow} & B(2) \cong \mathbb{Z} \\ & & \downarrow \tau_2 \\ \pi_1(U_{g+1}) & \xrightarrow[\chi(p_g)]{} & \Sigma_2 \cong \mathbb{Z}/2 , \end{array}$$

is then possible if and only if we can choose integers $d_1,\ldots,d_{g+1} \in \mathbb{Z}$ such that

$$\varphi'(c_i) = 2d_i + 1$$

and

$$\varphi'(c_1^2\cdot\ldots\cdot c_{g+1}^2) = \sum_{i=1}^{g+1} 2\cdot(2d_i+1) = 4\cdot(\sum_{i=1}^{g+1} d_i) + 2\cdot(g+1) = 0\,.$$

Clearly, this is possible if and only if $g+1$ is even, or equivalently, $g$ is odd.

According to Theorem 6.1 this proves, that the double covering map $p_g\colon T_g \to U_{g+1}$ is equivalent to a polynomial covering map if and only if $g$ is odd.

Thereby we get an alternative proof of Theorem 3.8.

## 7. Embedding finite covering maps into bundles of manifolds.

Throughout this section, X denotes a connected CW–complex of dimension $k \geq 1$, and M denotes a connected manifold of dimension $m \geq 2$.

The embedding criterion for polynomial covering maps leads naturally to questions concerning the existence of embeddings of finite covering maps into arbitrary bundles. Such problems have already been considered in several papers. References can be found in [36]. We shall present a new embedding theorem, which generalizes results for embeddings into vector bundles.

Let $\pi: E \to X$ be a finite covering map, and let $p: B \to X$ be a locally trivial bundle in the sense of Steenrod [13] with typical fibre M.

Definition 7.1. We say that $\pi: E \to X$ can be embedded into $p: B \to X$, if there is a continuous map (called an embedding) $h: E \to B$, which maps E homeomorphically onto its image $h(E)$ in B and commutes with the projections onto X,

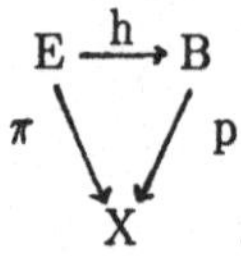


.

Recall, that a topological space Y is said to be q–connected $(q \geq 0)$ if it is pathwise connected and $\pi_i(Y) = 0$ for $i = 1,\dots,q$.

Our aim is to prove the following

Theorem 7.2. Let $p: B \to X$ be a locally trivial bundle in the sense of Steenrod over a connected CW–complex X of dimension $k \geq 1$ and with a $(k-1)$–connected manifold M of dimension $m > k \geq 1$ as typical fibre.

Then any finite covering map $\pi: E \to X$ embeds into $p: B \to X$.

Examples of bundles $p: B \to X$ to which Theorem 7.2 apply include vector bundles and sphere bundles of dimensions $m > k$. Theorem 7.2 is best possible in general, since the double covering map $\pi: S^k \to \mathbb{R}P^k$ does not embed into the trivial bundle $p: \mathbb{R}P^k \times \mathbb{R}^k \to \mathbb{R}P^k$. This follows easily by an application of the Borsuk–Ulam theorem, just as in the proof of Theorem 3.7.

The proof of Theorem 7.2 uses a method of Dold [25], which transforms the embedding problem into a corresponding section problem for a suitably defined locally trivial fibration.

First we shall define the appropriate locally trivial fibration.

Let $\pi: E \to X$ be an n–fold covering map, $n \geq 2$, and let $p: B \to X$ be the bundle into which we shall embed $\pi$.

For each point $x \in X$, denote by $E_x$ and $B_x$ the fibre over $x \in X$ of $\pi$, respectively $p$. Let $\mathrm{Emb}(E_x, B_x)$ denote the space of embeddings of $E_x$ into $B_x$. By choosing an ordering of the $n$ points in $E_x$ and identifying $B_x$ with $M$, we can identify $\mathrm{Emb}(E_x, B_x)$ with the configuration space $F_n(M)$. We can – and shall in the following – consider a configuration $c \in F_n(M)$ as an embedding $c: \{1,\dots,n\} \to M$.

Define the set

$$\mathrm{Emb}(\pi, p) = \coprod_{x \in X} \mathrm{Emb}(E_x, B_x) \quad \text{(disjoint union)}$$

and let

$$p^{\pi}: \mathrm{Emb}(\pi, p) \to X$$

be the map, which projects $\mathrm{Emb}(E_x, B_x)$ onto $x \in X$.

We shall equip $\mathrm{Emb}(\pi, p)$ with a topology, such that $p^{\pi}$ is a locally trivial fibration. We do this by describing what will eventually be local trivializations of $p^{\pi}$.

Let $U \subseteq X$ be an (open) set in $X$ over which both $\pi$ and $p$ are trivial, and let

$$\begin{array}{ccc} \pi^{-1}(U) & \xrightarrow{\ \xi\ } & U \times \{1,\dots,n\} \\ & {\scriptstyle \pi} \searrow \quad \swarrow {\scriptstyle \mathrm{proj}_1} & \\ & U & \end{array} \qquad \text{and} \qquad \begin{array}{ccc} U \times M & \xrightarrow{\ \varphi\ } & p^{-1}(U) \\ & {\scriptstyle \mathrm{proj}_1} \searrow \quad \swarrow {\scriptstyle p} & \\ & U & \end{array}$$

be local trivializations.

For each $x \in U$, let $\xi_x: E_x \to \{1,\dots,n\}$ and $\varphi_x: M \to B_x$ be the homeomorphisms on fibres defined by the local trivializations $\xi$ and $\varphi$ respectively.

Define a corresponding local trivialization $\varphi^{\xi}$ of $p^{\pi}$ over $U$,

$$\begin{array}{ccc} U \times F_n(M) & \xrightarrow{\ \varphi^{\xi}\ } & (p^{\pi})^{-1}(U) \\ & {\scriptstyle \mathrm{proj}_1} \searrow \quad \swarrow {\scriptstyle p^{\pi}} & \\ & U & \end{array} ,$$

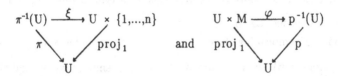

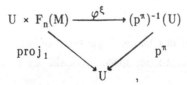

by associating to $(x, c) \in U \times F_n(M)$ the embedding $\varphi^\xi(x, c)\colon E_x \to B_x$ defined as the composition $\varphi^\xi(x, c) = \varphi_x \circ c \circ \xi_x$.

If the local trivialization $\psi^\eta$ of $p^\pi$ over $V$,

$$\begin{array}{ccc} V \times F_n(M) & \xrightarrow{\psi^\eta} & (p^\pi)^{-1}(V) \\ & {\scriptstyle proj_1} \searrow \quad \swarrow {\scriptstyle p^\pi} & \\ & V & , \end{array}$$

is defined similarly from local trivializations $\eta$ and $\psi$ of $\pi$, respectively $p$, over the (open) set $V \subseteq X$, then it is relatively easy to prove that

$$(\psi^\eta)^{-1} \circ \varphi^\xi\colon (U \cap V) \times F_n(M) \to (U \cap V) \times F_n(M)$$

is a homeomorphism.

Granted this information, it is easy to prove

Lemma 7.3. The set $\mathrm{Emb}(\pi, p)$ can be given a unique topology, such that $p^\pi\colon \mathrm{Emb}(\pi, p) \to X$ is a locally trivial fibration with typical fibre $F_n(M)$ and local trivializations $\varphi^\xi$ as defined above.

Furthermore, an embedding $h\colon E \to B$ of $\pi$ into $p$ induces a (continuous) section $s_h\colon X \to \mathrm{Emb}(\pi, p)$ of $p^\pi$, and conversely. The embedding $h$ and the section $s_h$ are related by the formula $s_h(x)(e_x) = h(e_x)$ for $x \in X$, $e_x \in E_x$.

We shall also need the following

Lemma 7.4. Suppose that $M$ is a $(k-1)$ – connected manifold of dimension $m > k \geq 1$. Then the configuration space $F_n(M)$ is $(k-1)$ – connected for any integer $n \geq 2$.

Proof. As already observed in Chapter I, §2, the configuration space $F_n(M)$ can be constructed from the n–fold product space $M\times\ldots\times M$ by removing the finitely many submanifolds

$$V_{ij} = \{(q_1,\ldots,q_n) \in M\times\ldots\times M \mid q_i = q_j\}$$

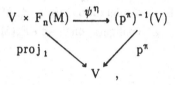

corresponding to pairs $(i, j)$ with $1 \leq i < j \leq n$, i.e.

$$F_n(M) = M \times \ldots \times M \setminus \bigcup_{1 \leq i < j \leq n} V_{ij} .$$

Since $V_{ij}$ is a submanifold of codimension $m$ in $M \times \ldots \times M$, any homotopy in dimensions $\leq m - 1$ can be pushed off $\bigcup_{1 \leq i < j \leq n} V_{ij}$ by elementary transversality theory. [The reader may wish to consult Birman ([24], in particular p. 43) for $M$ a smooth manifold and Kirby and Siebenmann ([6], Essay III, §1) for transversality theory in the topological category.] The homotopy groups of $F_n(M)$ in dimensions $\leq m - 2$ are therefore isomorphic to the corresponding homotopy groups of $M \times \ldots \times M$. Since $M$ is $(k - 1)$ – connected and $0 \leq k - 1 \leq m - 2$, we conclude that $F_n(M)$ is $(k - 1)$ – connected. This proves Lemma 7.4.

Now we are ready for the

Proof of Theorem 7.2. According to Lemma 7.3, we have only to prove that the locally trivial fibration $p^\pi$: Emb$(\pi, p) \to X$ has a section. Since both $\pi$ and $p$ are trivial over the cells of $X$ ([13], Corollary 11.6, p. 53), the fibration $p^\pi$ is trivial over the cells of $X$. By Lemma 7.4, the typical fibre $F_n(M)$ of $p^\pi$ is $(k - 1)$ – connected. Since $X$ is $k$–dimensional, the existence of a section for $p^\pi$ follows then by elementary obstruction theory ([13], Corollary 29.3, p. 149). This proves Theorem 7.2.

## EXERCISES

1. ([33], Theorem 8.3). Prove that every finite covering map onto the unit circle $S^1$ is polynomial and provide an explicit simple Weierstrass polynomial for each such covering map.

2. Let $T = S^1 \times S^1$ denote the 2–dimensional torus with coordinates $(x_1, x_2) \in S^1 \times S^1$, i.e. $x_1$ and $x_2$ are complex numbers of modulus $|x_1| = |x_2| = 1$. Prove that every finite connected covering space of $T$ is polynomial and provide an explicit simple Weierstrass polynomial for each such covering space.

3. Let $\delta(a_1,\ldots,a_n)$ denote the discriminant of the complex polynomial $P(z) = z^n + \sum_{i=1}^{n} a_i z^{n-i}$. Prove that for

$$n = 2: \quad \delta(a_1, a_2) = a_1^2 - 4a_2$$

$$n = 3: \quad \delta(0, a_2, a_3) = -4a_2^3 - 27a_3^2 .$$

4. Prove that the canonical 2–fold polynomial covering map $\pi^2: E^2 \to B^2$ can be identified with the 2–fold covering map $p_2: S^1 \to S^1$, defined by $p_2(z) = z^2$.

5. Prove Corollary 3.3 without reference to the pull–back construction, just using the path lifting property for covering maps.

6. Prove the Borsuk–Ulam theorem for orientable surfaces of even genus stated in Lemma 3.10 using characteristic homomorphisms for covering maps.

7. Describe all finite covering maps onto the product space $S^1 \times \mathbb{R}P^2$ and determine which ones are polynomial.

8. Let $\pi_n: S^n \to \mathbb{R}P^n$, $n \geq 1$, be the standard double covering of the real projective n–space $\mathbb{R}P^n$ by the n–sphere $S^n$. There are natural inclusions,

$$\begin{array}{ccccccccccc}
S^1 & = & S^1 & \subset & S^2 & \subset \ldots\ldots \subset & S^n & \subset \ldots\ldots \subset & S^\infty \\
{\scriptstyle p_2}\downarrow & & {\scriptstyle \pi_1}\downarrow & & {\scriptstyle \pi_2}\downarrow & & {\scriptstyle \pi_n}\downarrow & & {\scriptstyle \pi_\infty}\downarrow \\
S^1 & = & \mathbb{R}P^1 & \subset & \mathbb{R}P^2 & \subset \ldots\ldots \subset & \mathbb{R}P^n & \subset \ldots\ldots \subset & \mathbb{R}P^\infty .
\end{array}$$

In this diagram, $\pi_\infty: S^\infty \to \mathbb{R}P^\infty$ is the union of the coverings $\pi_n: S^n \to \mathbb{R}P^n$ for $n \geq 1$, and $S^\infty$ and $\mathbb{R}P^\infty$ are equipped with the weak topologies. At the bottom, $\pi_1: S^1 \to \mathbb{R}P^1$ can clearly be identified with the double covering map $p_2: S^1 \to S^1$, defined by $p_2(z) = z^2$.

It is well known that $\pi_\infty: S^\infty \to \mathbb{R}P^\infty$ classifies all 2–fold covering maps, since $S^\infty$ is contractible. In other words: Every 2–fold covering map $\pi: E \to X$ is induced from $\pi_\infty: S^\infty \to \mathbb{R}P^\infty$ via pull–back along a map $f: X \to \mathbb{R}P^\infty$. The Stiefel–Whitney class of $\pi: E \to X$ can be defined as the cohomology class $w_1(\pi) \in H^1(X; \mathbb{Z}/2)$ represented by the induced homomorphism $f_*: \pi_1(X) \to \mathbb{Z}/2$.

Let $\pi: E \to X$ be a 2–fold covering map with classifying map $f: X \to \mathbb{R}P^\infty$.

(i) Prove that the characteristic homomorphism of $\pi$ is given by $\chi(\pi) = \tau \circ f_*$, where $\tau: \pi_1(\mathbb{R}P^\infty) \to \mathbb{Z}/2$ is the boundary homomorphism (in fact an isomorphism) in $\pi_\infty: S^\infty \to \mathbb{R}P^\infty$.

(ii) Prove that the permutation homomorphism $\tau_2: B(2) \to \Sigma_2$ can be identified with the homomorphism $i_*: \pi_1(S^1) \to \pi_1(\mathbb{R}P^\infty)$ induced by the inclusion map $i: S^1 \to \mathbb{R}P^\infty$.

(iii) Prove that the 2–fold covering map $\pi: E \to X$ is polynomial if and only if the classifying map $f: X \to \mathbb{R}P^\infty$ factors through the inclusion map $i: S^1 \to \mathbb{R}P^\infty$.

The coefficient sequence $0 \to \mathbb{Z} \xrightarrow{\cdot 2} \mathbb{Z} \to \mathbb{Z}/2 \to 0$ induces an exact sequence in cohomology

$$H^1(X; \mathbb{Z}) \xrightarrow{\cdot 2} H^1(X; \mathbb{Z}) \xrightarrow{r} H^1(X; \mathbb{Z}/2) \to H^2(X; \mathbb{Z}) \to \dots ,$$

where $r$ is reduction of coefficients modulo 2.

(iv) ([33], Remark 7.5). Prove that the 2–fold covering map $\pi: E \to X$ is polynomial if and only if its Stiefel–Whitney class $w_1(\pi) \in H^1(X; \mathbb{Z}/2)$ is reduction modulo 2 of an integral class in $H^1(X; \mathbb{Z})$.

9. Let $\pi: E \to X$ be an n–fold polynomial covering map embedded into the trivial complex line bundle $X \times \mathbb{C}$ over $X$ such that $E \cap (X \times \{0\}) = \emptyset$. Show that

$$\pi_1((X \times \mathbb{C})\backslash E) = \mathbb{F}_n \ltimes \pi_1(X) ,$$

where the semi–direct product $\mathbb{F}_n \ltimes \pi_1(X)$ is w.r.t. the action

$$\pi_1(X) \xrightarrow{a_*} \pi_1(B^n) \subset \mathrm{Aut}(\mathbb{F}_n)$$

induced by the coefficient map $a: X \to B^n$ of $\pi$ .

Chapter IV

# ALGEBRA AND TOPOLOGY OF WEIERSTRASS POLYNOMIALS

Throughout this chapter, X denotes a connected and locally pathwise connected topological space with the homotopy type of a CW–complex. Let C(X) denote the ring of complex valued, continuous functions on X . A Weierstrass polynomial P(x, z) over X can then be viewed as an element in the polynomial ring C(X)[z] .

A natural first question concerning the algebra of a Weierstrass polynomial P(x, z) over X is to ask whether it has a root over C(X) , in other words, whether there exists a continuous function $\lambda: X \to \mathbb{C}$ such that $P(x, \lambda(x)) = 0$ . In §1, we shall present the basic elements of work of Gorin and Lin [32] containing necessary and sufficient conditions that a space X has to satisfy in order that every simple Weierstrass polynomial P(x, z) of degree $n \geq 2$ over X splits completely into linear factors over C(X) . Associated with a simple Weierstrass polynomial P(x, z) over X we have the polynomial covering map $\pi: E \to X$ . Complete solvability of the equation $P(x, z) = 0$ is equivalent to triviality of $\pi: E \to X$ . This is an example of the connections between the algebra of the simple Weierstrass polynomial P(x, z) on the one hand and the topology of the polynomial covering map $\pi: E \to X$ on the other hand.

For an arbitrary covering map $\pi: E \to X$ , there is an induced monomorphism of rings $\pi^*: C(X) \to C(E)$ . Thereby we can consider C(E) as a ring extension of C(X) , or as a C(X)–algebra. Duchamp and Hain [26] have characterized an n–fold polynomial covering map as an n–fold covering map $\pi: E \to X$ for which the ring extension C(E) of C(X) has a primitive, i.e. there is a function $f \in C(E)$ such that $1, f, \dots, f^{n-1}$ generate C(E) as a C(X)–module. We present their proof in §2. For a finite covering map $\pi: E \to X$ onto a compact space X , we prove in §3 that the C(X)–algebra C(E) characterizes the equivalence class of the covering map, and hence we call it the characteristic algebra of $\pi$ , [37]. Using the notion of a primitive, it is not difficult to prove that the characteristic algebra of the polynomial covering map $\pi: E \to X$ associated with the simple Weierstrass polynomial P(x, z) is the quotient algebra C(X)[z]/(P(x, z)) , where (P(x, z)) denotes the principal ideal in C(X)[z] generated by P(x, z) .

In §4, we explore how certain algebraic properties of the Weierstrass polynomial $P(x, z)$ over $X$ are reflected in algebraic properties of the $C(X)$–algebra $C(X)[z]/(P(x, z))$. Via the $C(X)$–algebra $C(X)[z]/(P(x, z))$, the algebra of the simple Weierstrass polynomial $P(x, z)$ is intimately related to the topology of the polynomial covering map $\pi: E \to X$. Applications of this are given in §5.

<u>1. Complete solvability of equations defined by simple Weierstrass polynomials.</u>

Consider a simple Weierstrass polynomial of degree $n \geq 2$ over $X$,

$$P_a(x, z) = z^n + \sum_{i=1}^{n} a_i(x)\, z^{n-i} .$$

Along with the coefficient map $a: X \to B^n$ for $P_a(x, z)$ we have as usual the root map $\alpha: X \to C_n(\mathbb{C})$ defined by the commutative diagram

$$\begin{array}{ccc} & \overset{\alpha}{\nearrow} & C_n(\mathbb{C}) \\ X & & \big\downarrow \Phi \\ & \underset{a}{\searrow} & B^n \end{array} .$$

Following Gorin and Lin [32], we make the

<u>Definition 1.1.</u> The equation

$$P_a(x, z) = z^n + \sum_{i=1}^{n} a_i(x)\, z^{n-1} = 0$$

is said to be <u>completely solvable</u> on $X$, if there exist $n$ continuous, complex valued functions $\lambda_1,\ldots,\lambda_n: X \to \mathbb{C}$ on $X$, such that

$$z^n + \sum_{i=1}^{n} a_i(x)\, z^{n-i} = \prod_{i=1}^{n} (z - \lambda_i(x)) .$$

In other words: The equation $P_a(x, z) = 0$ is completely solvable on $X$ if it admits $n$ globally defined, continuous solutions.

If we think of $P_a(x, z)$ as an element in the polynomial ring $C(X)[z]$, then complete solvability of $P_a(x, z) = 0$ is equivalent to saying that $P_a(x, z)$ splits completely into linear factors in $C(X)[z]$.

Recall that we have the principal $\Sigma_n$–bundle of configuration spaces $p_n: F_n(\mathbb{C}) \to C_n(\mathbb{C})$, which defines the braid group sequence

$$1 \to H(n) \xrightarrow{\rho_n} B(n) \xrightarrow{\tau_n} \Sigma_n \to 1 .$$

The connections between complete solvability of equations defined by simple Weierstrass polynomials, topology of polynomial covering maps, and homomorphisms into the Artin braid groups are laid down in the following

Theorem 1.2. Let $P_a(x, z) = z^n + \sum_{i=1}^{n} a_i(x)\, z^{n-i}$ be a simple Weierstrass polynomial of degree $n \geq 2$ over $X$ with root map $\alpha: X \to C_n(\mathbb{C})$. Then the following statements are equivalent:

(1) The equation $P_a(x, z) = 0$ is completely solvable on $X$.

(2) The polynomial covering map $\pi_a: E_a \to X$ associated with $P_a(x, z)$ is trivial.

(3) The image of the induced homomorphism $\alpha_*: \pi_1(X) \to B(n)$ is contained in $H(n)$.

Proof. A trivialization of the n–fold polynomial covering map $\pi_a: E_a \to X$,

$$\begin{array}{ccc} X \times \{1,\dots,n\} & \xrightarrow{\ h\ } & E_a \subset X \times \mathbb{C} \\ & {\scriptstyle \mathrm{proj}_1} \searrow \quad \swarrow {\scriptstyle \pi_a} & \\ & X & \end{array},$$

has the form $h(x, i) = (x, \lambda_i(x))$ for $n$ pointwise different, continuous functions $\lambda_1,\dots,\lambda_n: X \to \mathbb{C}$ solving the equation $P_a(x, z) = 0$. Conversely, it is clear that such functions $\lambda_1,\dots,\lambda_n: X \to \mathbb{C}$ provide a trivialization of $\pi_a: E_a \to X$ by the definition $h(x, i) = (x, \lambda_i(x))$ for $x \in X$ and $i = 1,\dots,n$. This proves the equivalence of statements (1) and (2).

The equivalence of statements (2) and (3) was proved in Theorem III.6.4.

This proves Theorem 1.2.

Gorin and Lin are in particular interested in finding necessary and sufficient conditions that a space $X$ has to satisfy in order that all equations defined by simple Weierstrass polynomials over $X$ are completely solvable.

From Theorem 1.2 we get immediately the

Corollary 1.3. For a space $X$ and an arbitrary degree $n \geq 2$, the following statements are equivalent:

(1) Every equation $P(x, z) = 0$ defined by a simple Weierstrass polynomial $P(x, z)$ of degree $n$ over $X$ is completely solvable.

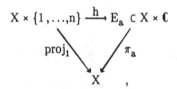

(2) Every n–fold polynomial covering map $\pi\colon E \to X$ onto $X$ is trivial.
(3) The image of every homomorphism $\varphi\colon \pi_1(X) \to B(n)$ is contained in $H(n)$.

For spaces $X$ with finitely generated first homology group, in particular finite complexes, we can do better.

Theorem 1.4. Suppose that the first homology group of $X$ with integer coefficients $H_1(X; \mathbb{Z})$ is finitely generated, and let $n \geq 2$ be an arbitrary degree. Then the following statements are equivalent:

(1) Every equation $P(x, z) = 0$ defined by a simple Weierstrass polynomial $P(x, z)$ of degree $n$ over $X$ is completely solvable.
(2) Every n–fold polynomial covering map $\pi\colon E \to X$ onto $X$ is trivial.
(3) Every homomorphism $\varphi\colon \pi_1(X) \to B(n)$ is trivial.

To prove this theorem we need some preparations.

First we shall describe a normal series for the coloured braid group $H(n)$.

Let $F_{m,n} = F_n(\mathbb{C}\backslash Q_m)$ denote the configuration space of $n$ ordered points in $\mathbb{C}$ avoiding $m$ points $Q_m = \{q_1,\ldots,q_m\} \subset \mathbb{C}$.

From Theorem I.2.1, or the proof of Corollary I.2.3, we have the diagram

$$\begin{array}{ccccccccccc}
\mathbb{C}\backslash Q_{n-1} = F_{n-1,1} & \to & F_{n-2,2} & \to \cdots \to & F_{2,n-2} & \to & F_{1,n-1} & \to & F_{0,n} = F_n(\mathbb{C}) \\
 & & \downarrow & & \downarrow & & \downarrow & & \downarrow \\
 & & F_{n-2,1} & & F_{2,1} & & F_{1,1} & & F_{0,1} \\
 & & \| & & \| & & \| & & \| \\
 & & \mathbb{C}\backslash Q_{n-2} & & \mathbb{C}\backslash Q_2 & & \mathbb{C}\backslash Q_1 & & \mathbb{C}
\end{array}$$

in which the vertical maps are the fibrations in Theorem I.2.1, and the horizontal maps are the inclusions of the corresponding fibres in the total spaces of the fibrations.

Put $Q_0 = \emptyset$ and $Q_k = \{1,\ldots,k\}$ for $1 \leq k \leq n-1$. Choose base points in the spaces in the diagram as follows:

$\bar{c}_k = (k+1,\ldots,n) \in F_{k,n-k}$ for $0 \leq k \leq n-1$
$(k+1) \in \mathbb{C}\backslash Q_k$ for $0 \leq k \leq n-1$.

For $0 \leq k \leq n-1$ define the group

$N_{n,k} = \pi_1(F_{k,n-k}, \bar{c}_k)$.

Since all homotopy groups of the spaces involved vanish in dimensions $\geq 2$, the homotopy sequence of the fibration

$$F_{k+1,n-k-1} \to F_{k,n-k} \to \mathbb{C}\backslash Q_k \ , \quad 0 \leq k \leq n-2 \ ,$$

reduces to the short exact sequence

$$1 \to N_{n,k+1} \to N_{n,k} \to \mathbb{F}_k \to 1 \ ,$$

where $\mathbb{F}_k$ is a free group on $k$ generators. Hence we can think of $N_{n,k+1}$ as a normal subgroup of $N_{n,k}$ with factor group a free group on $k$ generators.

Thereby we get the desired normal series for $H(n)$ as recorded in the following

Lemma 1.5. The coloured braid group $H(n)$ admits a normal series

$$H(n) = N_{n,0} \supset N_{n,1} \supset \ldots\ldots \supset N_{n,n-1} \supset N_{n,n} = \{1\}$$

in which each factor group $N_{n,k}/N_{n,k+1}$ is a free group on $k$ generators.

In order that a space $X$ has the property that every equation defined by a Weierstrass polynomial of degree $n \geq 2$ over $X$ has at least one continuous solution, it is a necessary condition that the first cohomology group of $X$ with integer coefficients $H^1(X; \mathbb{Z})$ is divisible by $n$. This is an immediate consequence of the following

Lemma 1.6. For a space $X$ and an arbitrary degree $n \geq 2$, the following statements are equivalent:

(1) Every equation of the type $z^n - q(x) = 0$, where $q: X \to \mathbb{C}\backslash\{0\}$ is a nonzero, complex valued, continuous function, admits a continuous solution $\lambda: X \to \mathbb{C}\backslash\{0\}$ (and hence is completely solvable).

(2) The degree $n$ divides $H^1(X; \mathbb{Z})$.

Proof. Consider the unit circle $S^1$ as the set of complex numbers $z \in \mathbb{C}$ of modulus $|z| = 1$, and define the n–fold covering map $p_n: S^1 \to S^1$ by $p_n(z) = z^n$.

Since the positive, real valued, continuous function $|q|: X \to \mathbb{R}_+$ defined by $|q|(x) = |q(x)|$ trivially has a continuous n'th root, we can restrict the attention to functions $q: X \to S^1$ in statement (1).

Observe now, that an equation $z^n - q(x) = 0$ for a continuous function $q\colon X \to S^1$ has a continuous solution $\lambda\colon X \to S^1$ if and only if $q\colon X \to S^1$ admits the lifting $\lambda\colon X \to S^1$ over $p_n$,

$$\begin{array}{ccc} & & S^1 \\ & \overset{\lambda}{\nearrow} & \Big\downarrow p_n \\ X & \xrightarrow[q]{} & S^1 \end{array} .$$

By general covering space theory, such a lifting of $q$ exists if and only if $q_*(\pi_1(X)) \subseteq (p_n)_*(\pi_1(S^1))$.

Under the identification $\pi_1(S^1) \cong \mathbb{Z}$, it is clear that $(p_n)_*\colon \pi_1(S^1) \to \pi_1(S^1)$ corresponds to multiplication by $n$. Therefore $q_*(\pi_1(X)) \subseteq (p_n)_*(\pi_1(S^1))$ if and only if there exists a homomorphism $\psi\colon \pi_1(X) \to \pi_1(S^1)$ such that $q_* = n\,\psi$.

There are well known isomorphisms

$$\begin{aligned} H^1(X;\mathbb{Z}) &\cong \operatorname{Hom}(H_1(X;\mathbb{Z}),\mathbb{Z}) \\ &\cong \operatorname{Hom}(\pi_1(X)/[\pi_1(X),\pi_1(X)],\mathbb{Z}) \\ &\cong \operatorname{Hom}(\pi_1(X),\mathbb{Z}), \end{aligned}$$

where $[\pi_1(X), \pi_1(X)]$ denotes the commutator subgroup of $\pi_1(X)$.

Since $S^1$ is an Eilenberg–MacLane space of type $(\mathbb{Z}, 1)$, it is well known that every homomorphism $\varphi\colon \pi_1(X) \to \mathbb{Z}$ is induced by a map $q\colon X \to S^1$ and hence that every element in $H^1(X;\mathbb{Z})$ is represented by an induced homomorphism $q_*\colon \pi_1(X) \to \pi_1(S^1)$.

We conclude, that every $q\colon X \to S^1$ admits a lifting $\lambda\colon X \to S^1$ over $p_n$ if and only if every induced homomorphism $q_*$ satisfies an equation $q_* = n\,\psi$ for a homomorphism $\psi\colon \pi_1(X) \to \pi_1(S^1)$, or equivalently, if and only if every element in $H^1(X;\mathbb{Z})$ is divisible by $n$.

This proves Lemma 1.6.

Now we are ready for the

Proof of Theorem 1.4. Most of the implications have already been proved in Corollary 1.3. It clearly suffices to prove that (1) ⇒ (3). Assume therefore that (1) is satisfied.

Since $H_1(X; \mathbb{Z})$ is finitely generated, it follows that $H^1(X; \mathbb{Z}) \cong \mathrm{Hom}(H_1(X; \mathbb{Z}), \mathbb{Z})$ is a free abelian group of finite rank.

Since (1) is satisfied, it follows by Lemma 1.6 that $H^1(X; \mathbb{Z})$ is divisible by $n$. Consequently, $H^1(X; \mathbb{Z}) = 0$ since it is free abelian.

Given a homomorphism $\varphi\colon \pi_1(X) \to B(n)$. By Corollary 1.3 we know that actually $\varphi\colon \pi_1(X) \to H(n)$. If $\varphi\colon \pi_1(X) \to H(n)$ were nontrivial, we could construct a nontrivial homomorphism $\psi\colon \pi_1(X) \to \mathbb{Z}$, since $H(n)$ admits the normal series with free factor groups described in Lemma 1.5, and since a subgroup of a free group is itself free. This would contradict that $\mathrm{Hom}(\pi_1(X), \mathbb{Z}) \cong H^1(X; \mathbb{Z}) = 0$. We conclude that $\varphi$ is trivial. This proves the implication (1) ⇒ (3).

As already remarked this finishes the proof of Theorem 1.4.

Using the tools developed in this section we can also prove the remarks immediately preceding Example III.4.2. We state this formally as

Theorem 1.7. Let $n \geq 2$ be an arbitrary degree. Then the following statements are equivalent:

(1) The first cohomology group $H^1(X; \mathbb{Z}) = 0$.
(2) Any coefficient map $a\colon X \to B^n$, for which the associated n–fold polynomial covering map $\pi_a\colon E_a \to X$ is trivial, is homotopic to a constant map.

Or equivalently: If and only if $H^1(X; \mathbb{Z}) = 0$ does $X$ not allow a homotopically nontrivial map $a\colon X \to B^n$ with corresponding trivial polynomial covering map $\pi_a\colon E_a \to X$.

Proof. (1) ⇒ (2). Suppose $\pi_a\colon E_a \to X$ is trivial. Then by Theorem 1.2, the image of $\alpha_*\colon \pi_1(X) \to B(n)$ is contained in $H(n)$, i.e. $\alpha_*\colon \pi_1(X) \to H(n)$. If $\alpha_*$ was nontrivial, then we could define a nontrivial homomorphism $\psi\colon \pi_1(X) \to \mathbb{Z}$ as in the proof of Theorem 1.4. But this would contradict that $\mathrm{Hom}(\pi_1(X), \mathbb{Z}) \cong H^1(X; \mathbb{Z}) = 0$. Therefore $\alpha_*\colon \pi_1(X) \to B(n)$ is trivial. Since $C_n(\mathbb{C})$ is an Eilenberg–MacLane space of type $(B(n), 1)$ it follows that $\alpha\colon X \to C_n(\mathbb{C})$, and hence $a\colon X \to B^n$, is homotopically trivial.

(2) ⇒ (1) . The proof is by contradiction. Suppose that $\mathrm{Hom}(\pi_1(X), \mathbb{Z}) \cong H^1(X; \mathbb{Z}) \neq 0$ . Since $H(n)$ admits a normal series with free factor groups, we can construct a nontrivial homomorphism $\varphi: \pi_1(X) \to H(n)$ . Since $C_n(\mathbb{C})$ is an Eilenberg-MacLane space of type $(B(n), 1)$ , this homomorphism can be realized by a map $\alpha: X \to C_n(\mathbb{C})$ , which defines an n–fold polynomial covering map $\pi_a: E_a \to X$ with coefficient map $a = \Phi \circ \alpha$ . The map $a$ is then homotopically nontrivial, whereas $\pi_a: E_a \to X$ is trivial by Theorem 1.2. Contradiction.

This proves Theorem 1.7.

## 2. Primitives for extensions of rings of continuous functions.

For any topological space $Y$, we denote by $C(Y)$ the ring of complex valued, continuous functions on $Y$ with its usual pointwise defined algebraic structure. A continuous map $h: Y_1 \to Y_2$ induces a ring homomorphism $h^*: C(Y_2) \to C(Y_1)$ by the definition $h^*(c) = c \circ h$ for any continuous function $c \in C(Y_2)$.

Let $\pi: E \to X$ be an n–fold covering map. Then we have the induced homomorphism

$$\pi^*: C(X) \to C(E)\,, \text{ defined by } \pi^*(c) = c \circ \pi \text{ for } c \in C(X)\,.$$

Since $\pi$ is a covering map, it is easy to see that $\pi^*$ is a monomorphism. Via $\pi^*$ we can therefore consider $C(E)$ as a ring extension of $C(X)$, or a $C(X)$–module, or a $C(X)$–algebra. To be specific: If $f \in C(E)$ and $c \in C(X)$ we define the continuous function $c \cdot f \in C(E)$ by $c \cdot f(e) = (c \circ \pi)(e)\, f(e)$ for $e \in E$ using the product in $\mathbb{C}$ on the right hand side of the equation.

The $C(X)$–algebra $C(E)$ is commutative and it has an identity element, represented by the function with constant value 1.

For the rest of this section we assume that $X$ is a completely regular space [5], i.e.: $X$ is a Hausdorff space, and for any closed set $A$ in $X$ and any point $x_0 \in X\backslash A$ there exists a continuous function $c \in C(X)$ such that $c$ is $0$ on $A$ and $c(x_0) = 1$. It is then easy to prove, that if $E$ is a covering space of $X$, and $e_1, e_2 \in E$ are points in $E$ such that $g(e_1) = g(e_2)$ for all $g \in C(E)$, then $e_1 = e_2$.

In accordance with similar notions from the theory of fields, we make the

Definition 2.1. Let $\pi: E \to X$ be an n–fold covering map. A continuous function $f \in C(E)$ is called a primitive for the ring extension $C(E)$ of $C(X)$, if the set of functions $1, f, \dots, f^{n-1}$ is a basis for $C(E)$ as a $C(X)$–module, or in other words: For every $g \in C(E)$ there exist unique functions $c_0, c_1, \dots, c_{n-1} \in C(X)$ such that

$$g = c_0 + c_1 \cdot f + \dots + c_{n-1} \cdot f^{n-1}\,.$$

The existence of a primitive is equivalent to the existence of an embedding of the covering map into the trivial complex line bundle. This result is due to Duchamp and Hain [26] and is contained in the following

Theorem 2.2. Let $\pi: E \to X$ be an n-fold covering map. Then the following statements are equivalent for a continuous function $f: E \to \mathbb{C}$:

(1) The function $f \in C(E)$ is a primitive for the ring extension $C(E)$ of $C(X)$.

(2) The continuous function $f: E \to \mathbb{C}$ separates points in the fibres of $\pi$, i.e. $f(e_1) \neq f(e_2)$ for $e_1, e_2 \in E$ if $\pi(e_1) = \pi(e_2)$ and $e_1 \neq e_2$.

(3) The continuous map $h = (\pi, f): E \to X \times \mathbb{C}$ is an embedding of $\pi: E \to X$ into the trivial complex line bundle $\mathrm{proj}_1: X \times \mathbb{C} \to X$ over $X$.

Proof. The statements (2) and (3) are obviously equivalent, so we only have to prove that (1) and (2) are equivalent.

Suppose first, that $f \in C(E)$ is a primitive. Then obviously $f: E \to \mathbb{C}$ must separate points in the fibres, since if $f \in C(E)$ did not separate points in the fibres, no function $g \in C(E)$ would do, and that would be a contradiction.

Suppose next, that $f: E \to \mathbb{C}$ separates points in the fibres of $\pi$. Consider an arbitrary point $x \in X$, and let $\{e_1,\dots,e_n\} = \pi^{-1}(x)$ be the fibre of $\pi$ over $x \in X$.

Given a function $g \in C(E)$, we have to solve the following system of linear equations over $\mathbb{C}$ for $c_0(x), c_1(x),\dots,c_{n-1}(x)$,

$$g(e_1) = c_0(x) + c_1(x)\, f(e_1) + \dots + c_{n-1}(x)\, f^{n-1}(e_1)$$
$$g(e_2) = c_0(x) + c_1(x)\, f(e_2) + \dots + c_{n-1}(x)\, f^{n-1}(e_2)$$
$$\vdots$$
$$g(e_n) = c_0(x) + c_1(x)\, f(e_n) + \dots + c_{n-1}(x)\, f^{n-1}(e_n) .$$

The determinant of the system is the Vandermonde determinant

$$\prod_{1 \le j < i \le n} (f(e_i) - f(e_j)) ,$$

which is nonzero, since $f$ separates points in the fibres of $\pi$. Hence there is a unique solution $c_0(x), c_1(x),\dots,c_{n-1}(x)$ to the system of equations.

Since $\pi: E \to X$ is locally trivial, the functions $c_0, c_1,\dots,c_{n-1}: X \to \mathbb{C}$ so defined will be continuous functions, and by construction

$$g = c_0 + c_1 \cdot f + \dots + c_{n-1} \cdot f^{n-1} .$$

This proves that $f \in C(E)$ is a primitive for the ring extension $C(E)$ of $C(X)$ and hence finishes the proof of Theorem 2.2.

Using the embedding criterion (Theorem III.3.6) we conclude:

Corollary 2.3. An n–fold covering map $\pi: E \to X$ is equivalent to a polynomial covering map if and only if the ring extension $C(E)$ of $C(X)$ admits a primitive.

Suppose that $\pi: E \to X$ is an n–fold polynomial covering map. Then $E \subset X \times \mathbb{C}$, and we can define the continuous function $f: E \to \mathbb{C}$ by $f(x, z) = z$. Clearly, $f$ separates points in the fibres of $\pi$ and hence it is a primitive for the ring extension $C(E)$ of $C(X)$. Thereby we get the following

Corollary 2.4. If $\pi: E \to X$ is an n–fold polynomial covering map, then any continuous function $g: E \to \mathbb{C}$ has a unique decomposition

$$g(x, z) = c_0(x) + c_1(x)\, z + \dots + c_{n-1}(x)\, z^{n-1}$$

for some continuous functions $c_0, c_1, \dots, c_{n-1}: X \to \mathbb{C}$.

## 3. The characteristic algebra of a finite covering map.

The material in this section is taken from [37]. Throughout we shall work with finite covering maps $\pi: E \to X$ onto a fixed compact space $X$. Note that $E$ is then also compact. As usual we assume $X$ to be connected, but $E$ need not be connected.

In the following it is possible to extend the class of compact spaces to the wider class of realcompact spaces in the sense of Hewitt; see e.g. the book by Gillman and Jerison ([5], Chapter 8). The class of realcompact spaces includes among others the class of separable metric spaces. Again $E$ is realcompact if $X$ is realcompact.

Let $\pi: E \to X$ be a finite covering map of the appropriate type, and consider the $C(X)$–algebra $C(E)$ induced by $\pi$. We shall prove below that this $C(X)$–algebra determines the topological equivalence class of the covering map. Hence we call $C(E)$ the characteristic algebra over $C(X)$ of the covering map $\pi$, or the characteristic $C(X)$–algebra of $\pi$.

The basic theorem is

Theorem 3.1. Let $\pi_i: E_i \to X$, $i = 1, 2$, be finite covering maps onto the compact space $X$. Then we have:

An equivalence of covering maps,

$$\begin{array}{ccc} E_1 & \xrightarrow{h} & E_2 \\ & {}_{\pi_1}\searrow \quad \swarrow_{\pi_2} & \\ & X & \end{array} ,$$

induces an isomorphism of $C(X)$–algebras,

$$\varphi = h^*: C(E_2) \to C(E_1) ,$$

and conversely.

The proof of Theorem 3.1 uses a classical theorem of Gelfand and Kolmogoroff, the machinery behind which we first recall.

For a compact space $E$, consider the set of maximal ideals $\mathfrak{m}$ in the ring $C(E)$. Denote this set by $\Phi_{C(E)}$. On $\Phi_{C(E)}$ we can put a topology, in which the closed sets are given by

$$V(\mathfrak{a}) = \{\mathfrak{m} \in \Phi_{C(E)} \mid \mathfrak{a} \subseteq \mathfrak{m}\}$$

for the ideals $\mathfrak{a}$ in $C(E)$. This topology is called the Zariski topology, or in this context also the Stone topology, or the Gelfand topology on $\Phi_{C(E)}$. The set $\Phi_{C(E)}$ of maximal ideals in $C(E)$ equipped with the Zariski topology is called *the spectrum* or *the carrier space* of $C(E)$. It is easy to prove that any maximal ideal in $C(E)$ is a so-called *fixed ideal*, i.e. it has the form

$$\mathfrak{m}_e = \{f \in C(E) \mid f(e) = 0\} \quad \text{for a point } e \in E .$$

The crux of the matter is now that the map $E \to \Phi_{C(E)}$, which to $e \in E$ associates the maximal ideal $\mathfrak{m}_e \in \Phi_{C(E)}$, is a homeomorphism.

Using the above machinery, it is easy to prove that an isomorphism of rings $\varphi\colon C(E_2) \to C(E_1)$ is induced by a homeomorphism $h\colon E_1 \to E_2$. Thereby we get the classical

*Theorem* (Gelfand and Kolmogoroff). Let $E_1$ and $E_2$ be compact spaces. Then a homeomorphism $h\colon E_1 \to E_2$ induces an isomorphism of rings $\varphi = h^*\colon C(E_2) \to C(E_1)$, and conversely.

With more care, the theorem can be extended to realcompact spaces; see e.g. the book by Gillman and Jerison [5].

*Proof of Theorem 3.1.*

Suppose first, that we have an equivalence of covering maps onto $X$,

$$\begin{array}{ccc} E_1 & \xrightarrow{h} & E_2 \\ & {\scriptstyle \pi_1}\searrow \quad \swarrow{\scriptstyle \pi_2} & \\ & X & \end{array} .$$

Since the homeomorphism $h\colon E_1 \to E_2$ commutes with projections onto $X$, we have the following computation for every choice of $f \in C(E_2)$, $c \in C(X)$ and $e_1 \in E_1$:

$$\begin{aligned} h^*(c \cdot f)(e_1) &= (c \cdot f)(h(e_1)) \\ &= c(\pi_2(h(e_1)))\, f(h(e_1)) \\ &= c(\pi_1(e_1))\, f(h(e_1)) \\ &= (c \cdot h^*(f))(e_1) . \end{aligned}$$

This proves that $h^*(c \cdot f) = c \cdot h^*(f)$ and hence that $\varphi = h^*\colon C(E_2) \to C(E_1)$ is an isomorphism of C(X)–algebras.

Suppose then conversely, that we have an isomorphism of C(X)–algebras,

$$\varphi\colon C(E_2) \to C(E_1) .$$

In particular, $\varphi$ is then an isomorphism of rings and hence by the classical theorem of Gelfand and Kolmogoroff there is a homeomorphism $h\colon E_1 \to E_2$, such that $\varphi = h^*$, as ring isomorphisms.

We shall prove that $h\colon E_1 \to E_2$ is actually an equivalence of covering maps. To prove this, we only have to show that h commutes with projections onto X, or equivalently, that it preserves fibres.

The proof is by contradiction. Suppose that h does not preserve fibres, and let $e_1 \in E_1$ be a point such that $\pi_2(h(e_1)) \neq \pi_1(e_1)$. Since X is compact, in particular a normal space, we can choose a function $c \in C(X)$ such that $c(\pi_1(e_1)) = 1$ and $c(\pi_2(h(e_1))) = 0$. Since $h^* = \varphi$ is an isomorphism of C(X)–algebras, we have

$$\begin{aligned} c \circ (\pi_2 \circ h) &= (c \circ \pi_2) \circ h = h^*(c \circ \pi_2) \\ &= h^*(c \cdot (1 \circ \pi_2)) = c \cdot h^*(1 \circ \pi_2) \\ &= (c \circ \pi_1) \cdot (1 \circ \pi_2 \circ h) = c \circ \pi_1 , \end{aligned}$$

where $1 \in C(X)$ denotes the function on X with constant value 1 and the dot is multiplication in C(X)–algebras. Applying this to $e_1 \in E_1$, we get $c(\pi_2(h(e_1))) = c(\pi_1(e_1))$, which is obviously a contradiction. Therefore $h\colon E_1 \to E_2$ must preserve fibres, which proves that h is an equivalence of covering maps onto X.

This proves Theorem 3.1.

From Theorem 3.1 we get immediately

Corollary 3.2. A finite covering map $\pi\colon E \to X$ onto a compact space X is trivial if and only if the characteristic C(X)–algebra of $\pi$ is a free C(X)–algebra of finite rank.

Proof. We only have to observe that the characteristic $C(X)$–algebra of the trivial n–fold covering map $\mathrm{proj}_1: X \times \{1,...,n\} \to X$ is isomorphic as a $C(X)$–algebra to the free $C(X)$–algebra $C(X)\times...\times C(X)$ of rank $n$. Corollary 3.2 is then a special case of Theorem 3.1.

We turn then to polynomial covering maps. In this case we can identify the characteristic algebra with an algebra directly related to the Weierstrass polynomial describing the polynomial covering map in question.

Let $\pi: E \to X$ be a polynomial covering map associated with the simple Weierstrass polynomial $P(x, z)$ over $X$. Let $(P(x, z))$ denote the principal ideal in the polynomial ring $C(X)[z]$ generated by $P(x, z)$. The quotient ring $C(X)[z]/(P(x, z))$ has then a natural structure as a $C(X)$–algebra. This $C(X)$–algebra is isomorphic to the characteristic algebra of the covering map $\pi: E \to X$. This is the content of the following

Theorem 3.3. Let $\pi: E \to X$ be a polynomial covering map associated with the simple Weierstrass polynomial $P(x, z)$ over $X$. Then there is an isomorphism of $C(X)$–algebras,

$$\psi: C(X)[z]/(P(x, z)) \to C(E) ,$$

defined by mapping the residue class of the polynomial $S(x, z) \in C(X)[z]$ onto the restriction $S|E: E \to \mathbb{C}$ of $S$ to $E \subset X \times \mathbb{C}$.

Proof. Suppose that $P(x, z)$ is a simple Weierstrass polynomial of degree $n \geq 1$. By polynomial division, it follows then easily, that $C(X)[z]/(P(x, z))$ is freely generated by the polynomials $1, z,...,z^{n-1}$ as a $C(X)$–module. On the other hand, we know by Corollary 2.4, that also $C(E)$ is freely generated as a $C(X)$–module by the polynomials $1, z,...,z^{n-1}$ considered as functions on $E \subset X \times \mathbb{C}$. The map $\psi$ defined in Theorem 3.3 clearly preserves the basis $1, z,...,z^{n-1}$, and hence $\psi$ is an isomorphism of $C(X)$–modules. Since $\psi$ is obviously a $C(X)$–algebra homomorphism, it must be an isomorphism of $C(X)$–algebras. This proves Theorem 3.3.

Remark 3.4. Suppose $X$ is compact, and let $(x, z) \in X \times \mathbb{C}$ be a zero for the simple Weierstrass polynomial $P(x, z)$ over $X$, i.e. a point in the associated polynomial covering space $E$. The set of residue classes in $C(X)[z]/(P(x, z))$ of polynomials $Q(x, z) \in C(X)[z]$, which vanish at $(x, z)$, i.e. $Q(x, z) = 0$, is a maximal ideal, which we denote by $\mathfrak{m}_{(x,z)}$ . Using the isomorphism $\psi$ of C(X)–algebras from Theorem 3.3, it follows by Gelfand – Kolmogoroff theory that every maximal ideal in $C(X)[z]/(P(x,z))$ has the form $\mathfrak{m}_{(x,z)}$ for a point $(x, z) \in X \times \mathbb{C}$ with $P(x, z) = 0$ , and that the set of maximal ideals in $C(X)[z]/(P(x, z))$ equipped with the Zariski topology is homeomorphic to E .

Combining Theorem 3.1 and Theorem 3.3 we get

<u>Corollary 3.5.</u> Let $\pi_i\colon E_i \to X$ , $i = 1, 2$, be polynomial covering maps associated with the simple Weierstrass polynomials $P_i(x, z)$ over the compact space X . Then we have:

An equivalence of covering maps,

$$\begin{array}{ccc} E_1 & \xrightarrow{\ h\ } & E_2 \\ & {\scriptstyle\pi_1}\searrow \quad \swarrow{\scriptstyle\pi_2} & \\ & X & \end{array},$$

induces an isomorphism of C(X)–algebras ,

$$\varphi = h^*\colon C(X)[z]/(P_2(x, z)) \to C(X)[z]/(P_1(x, z)) ,$$

and conversely.

<u>Proof.</u> The commutative diagram

$$\begin{array}{ccc} C(X)[z]/(P_2(x, z)) & \xrightarrow{\varphi = h^*} & C(X)[z]/(P_1(x, z)) \\ {\scriptstyle\psi_2}\Big\downarrow\cong & & \cong\Big\downarrow{\scriptstyle\psi_1} \\ C(E_2) & \xrightarrow[\varphi = h^*]{} & C(E_1) \end{array}$$

defines the isomorphism $\varphi = h^*$ of C(X)–algebras in the upper arrow from the one in the lower arrow defined in Theorem 3.1.

We also get the following

Corollary 3.6. A finite covering map $\pi: E \to X$ onto a compact space $X$ is equivalent to a polynomial covering map if and only if the characteristic $C(X)$–algebra $C(E)$ of $\pi$ is isomorphic to a $C(X)$–algebra of the form $C(X)[z]/(P(x,z))$ for a simple Weierstrass polynomial $P(x, z)$ over $X$.

In the theory of commutative complex Banach algebras, the $C(X)$–algebra $C(X)[z]/(P(x, z))$ is called an extension of $C(X)$ of algebraic type, or of Arens – Hoffman type, since they were introduced first in 1956 in the paper [21] by these authors. There is an extensive literature on such algebras.

4. Weierstrass polynomials and characteristic algebras.

In this section we shall discuss how certain algebraic properties of a Weierstrass polynomial $P(x, z)$ over a topological space $X$ are reflected in algebraic properties of the $C(X)$–algebra $C(X)[z]/(P(x, z))$ . The material is based on [37].

First we recall some notions from algebra. Proofs of the results mentioned can be found in the books by DeMeyer and Ingraham [3] and R.S. Pierce [11].

Let $R$ denote a commutative ring and $S$ a commutative algebra over $R$ . We assume that both $R$ and $S$ have an identity element, indiscriminately denoted by $1$ .

An $R$–module $M$ is called projective, if $M$ is isomorphic as an $R$–module to a direct summand of some free $R$–module.

The algebra $S$ is called a separable $R$–algebra, if $S$ is projective as a bimodule over itself, i.e. as a module over the ring $S \otimes_R S$ . If $S$ is also projective as an $R$–module, then separability implies that $S$ is a finitely generated $R$–module. The latter is always the case if $R$ is a field, since $S$ is then a free module over $R$ , in particular a projective $R$–module.

If $R$ is a field, then a finite dimensional $R$–algebra $S$ is called separable in the classical sense, if for every field extension $K$ of $R$ , the $K$–algebra $S \otimes_R K$ obtained by extending the coefficients has Jacobson radical $J(S \otimes_R K) = 0$ , or equivalently, $S \otimes_R K$ is a semi–simple $S \otimes_R K$–module. Here the Jacobson radical $J(S \otimes_R K)$ is defined as the intersection of all maximal ideals in $S \otimes_R K$ , and as usual a module is called simple if it has no nontrivial proper submodules, and semi–simple if it is a direct sum of simple modules. We mention that a stunted polynomial algebra $R[z]/(f(z))$ defined by a monic polynomial $f(z) \in R[z]$ over the field $R$ is separable in the classical sense if and only if $f(z)$ has no multiple roots in any field extension $K$ of $R$ . The latter fact follows easily by observing that the maximal ideals in $(R[z]/(f(z))) \otimes_R K = K[z]/(f(z))$ are exactly the principal ideals generated by the irreducible factors of $f(z)$ over $K$ .

We need two fundamental results about separable algebras over commutative rings.

Theorem A. If $R$ is a field, then an $R$–algebra $S$ is separable if and only if it is separable in the classical sense and the dimension of $S$ as a vector space over $R$ is finite.

Theorem B. Suppose $S$ is a finitely generated algebra over the commutative ring $R$ . Then $S$ is separable if and only if $S/\mathfrak{m}S$ is a separable $(R/\mathfrak{m})$–algebra for any maximal ideal $\mathfrak{m}$ in $R$ .

For proofs of Theorem A and Theorem B see the book by DeMeyer and Ingraham [3], Theorem 2.5, p. 50, and Theorem 7.1, p. 72, respectively. According to Theorem B, separability of an algebra can be tested at the maximal ideals of the ring, and since $R/\mathfrak{m}$ is a field, we are back to separability in the classical sense by Theorem A.

With these preparations we now turn to our main concern, namely Weierstrass polynomials and their associated algebras.

Let $P(x, z) = z^n + \sum_{i=1}^{n} a_i(x) z^{n-i}$ be a Weierstrass polynomial over the topological space $X$. Taking the discriminant of $P(x, z)$ for each $x \in X$, see Chapter III, §2, we get a continuous function

$$D = D(x) = \delta(a_1(x),\dots,a_n(x)): X \to \mathbb{C} ,$$

which we call the discriminant function of $P(x, z)$.

Then we have

Theorem 4.1. For a Weierstrass polynomial $P(x, z)$ over a compact space $X$, the following statements are equivalent:

(1) $P(x, z)$ is a simple Weierstrass polynomial.

(2) The discriminant function $D(x) \neq 0$ for all $x \in X$.

(3) The discriminant function $D \in C(X)$ is invertible.

(4) The $C(X)$–algebra $C(X)[z]/(P(x, z))$ is separable.

Proof. The statements (1), (2) and (3) are obviously equivalent. So we only have to prove the equivalence of statements (1) and (4).

The algebra $C(X)[z]/(P(x, z))$ over $C(X)$ is clearly finitely generated. By Gelfand – Kolmogoroff theory we also know that any maximal ideal in $C(X)$ has the form $\mathfrak{m}_{x_0} = \{f \in C(X) \mid f(x_0) = 0\}$ for a point $x_0 \in X$, and that $C(X)/\mathfrak{m}_{x_0} \cong \mathbb{C}$. Then it follows by Theorem B stated above, that $C(X)[z]/(P(x, z))$ is a separable $C(X)$–algebra if and only if $(C(X)/\mathfrak{m}_{x_0})[z]/(P(x_0, z))$ is a separable $(C(X)/\mathfrak{m}_{x_0})$–algebra for every $x_0 \in X$, or equivalently, if and only if $\mathbb{C}[x]/(P(x_0, z))$ is a separable $\mathbb{C}$–algebra for every $x_0 \in X$. By Theorem A stated above this is equivalent to $P(x_0, z)$ having no multiple roots for every $x_0 \in X$, or in other words, that $P(x, z)$ is a simple Weierstrass polynomial.

This proves that the statements (1) and (4) are equivalent and hence the theorem.

Theorem 4.1 shows in particular, that the characteristic algebra $C(X)[z]/(P(x, z))$ for the polynomial covering map $\pi: E \to X$ associated with the simple Weierstrass polynomial $P(x, z)$ is a separable $C(X)$–algebra.

An idempotent in a ring $S$ with identity element $1 \in S$ is an element $e \in S$ such that $e^2 = e$. The elements $e = 0, 1$ are always idempotents. Rings where these elements are the only idempotents have particular interest.

A Weierstrass polynomial $P(x, z)$ over a connected topological space $X$ is called reducible, if it can be decomposed as a product $P(x, z) = P_1(x, z) \cdot P_2(x, z)$ of two nontrivial lower degree Weierstrass polynomials $P_1(x, z)$ and $P_2(x, z)$ over $X$. If $P(x, z)$ is not reducible it is called irreducible over $X$. We remark that the main point in the definition of reducibility of a Weierstrass polynomial over $X$ is that the coefficient functions must also be continuous functions on $X$. Furthermore, we remark that if a simple Weierstrass polynomial is reducible over $X$, then the factors will also be simple Weierstrass polynomials over $X$.

With these definitions we have the following

Theorem 4.2. Let $P(x, z)$ be a simple Weierstrass polynomial over the connected topological space $X$, and let $\pi: E \to X$ be the associated polynomial covering map. Then the following statements are equivalent:

(1) The Weierstrass polynomial $P(x, z)$ is irreducible over $X$.

(2) The polynomial covering space $E$ is connected.

(3) The $C(X)$–algebra $C(X)[z]/(P(x, z))$ has no other idempotents than 0 and 1.

Proof. We start out by proving the equivalence of the negations to the statements (1) and (2) and thereby the equivalence of (1) and (2).

First suppose that $P(x, z)$ is reducible, say that $P(x, z) = P_1(x, z) \cdot P_2(x, z)$. Then $P_1(x, z)$, respectively $P_2(x, z)$, will define a polynomial covering space $E_1$, respectively $E_2$, over $X$. It is clear that $E = E_1 \cup E_2$ provides a splitting of $E$ into non–void, open subsets of $E$. Hence $E$ is not connected.

Next suppose that $E$ is not connected. Since $X$ is connected, a connected component $E_1$ of $E$ defines a covering space over $X$. For each $x \in X$ we have

$$P(x, z) = z^n + \sum_{i=1}^{n} a_i(x)\, z^{n-i} = \prod_{z_x \in J_x} (z - z_x) \ ,$$

where $J_x = \{z_x \in \mathbb{C} \mid P(x, z_x) = 0\}$ . Obviously, the product of the first degree polynomials $z - z_x$ corresponding to roots $z_x \in \pi^{-1}(x) \cap E_1$ will define a simple Weierstrass polynomial $P_1(x, z)$ over $X$, when $x$ varies over $X$. Similarly, the product of the first degree polynomials corresponding to roots $z_x \notin \pi^{-1}(x) \cap E_1$ will define a simple Weierstrass polynomial $P_2(x, z)$ over $X$. Clearly $P(x, z) = P_1(x, z) \cdot P_2(x, z)$ and hence $P(x, z)$ is reducible.

Then we turn to the proof of the equivalence of statements (2) and (3).

First observe, that the idempotents in the algebra $C(X)[z]/(P(x, z))$ correspond to the idempotents in the algebra $C(E)$, since these algebras are isomorphic as $C(X)$–algebras by Theorem 3.3. On the other hand it is well known – and not difficult to prove directly – that $E$ is connected if and only if $C(E)$ has no other idempotents than $0$ and $1$. The equivalence of statements (2) and (3) follows.

This proves Theorem 4.2.

Let $P(x, z) = z^n + \sum_{i=1}^{n} a_i(x)\, z^{n-i}$ be a Weierstrass polynomial of degree $n \geq 1$ over $X$. By polynomial division it follows immediately that the $C(X)$–algebra $C(X)[z]/(P(x, z))$ is freely generated by the polynomials $1, z, \ldots, z^{n-1}$ as a module over $C(X)$. In particular, the algebra $C(X)[z]/(P(x, z))$ is therefore a projective $C(X)$–module.

We finish this section by mentioning, that for commutative separable algebras, which are projective as modules over the coefficient ring and whose only idempotents are $0$ and $1$, there is a suitable Galois theory. This is the main theme in DeMeyer and Ingraham ([3], Chapter III). By Theorem 4.1 and Theorem 4.2 this will then apply to the characteristic algebras of polynomial covering maps $\pi\colon E \to X$ onto compact, connected spaces $X$.

## 5. Some applications of characteristic algebras.

The results in this section are mainly related to the following general question: How much can we change a given simple Weierstrass polynomial without changing the topological equivalence class of the associated polynomial covering map?

The theory of characteristic algebras is well suited to address this question. In addition to proofs using characteristic algebras we offer however also direct proofs of the theorems. The material is taken from [35] and [37].

In the first application we prove that the term of degree $n-1$ in a simple Weierstrass polynomial of degree n can always be eliminated without changing the equivalence class of the associated polynomial covering map. The proof follows very classical lines.

Theorem 5.1. For every simple Weierstrass polynomial $P(x, z) = z^n + \sum_{i=1}^{n} a_i(x)\, z^{n-i}$ over the (compact) space $X$, there exists a simple Weierstrass polynomial $\tilde{P}(x, w) = w^n + \sum_{i=1}^{n} \tilde{a}_i(x)\, w^{n-i}$ over $X$ with $\tilde{a}_1(x) = 0$ for all $x \in X$, such that the associated polynomial covering maps are equivalent.

Proof. A (Using characteristic algebras). By a very classical algebraic procedure we can rewrite $P(x, z)$ in the form

$$P(x, z) = \left(z + \frac{a_1(x)}{n}\right)^n + \sum_{i=2}^{n} \tilde{a}_i(x) \left(z + \frac{a_1(x)}{n}\right)^{n-i} ,$$

for suitable continuous function $\tilde{a}_2,\ldots,\tilde{a}_n\colon X \to \mathbb{C}$. Put $\tilde{P}(x, w) = w^n + \sum_{i=2}^{n} \tilde{a}_i(x)\, w^{n-i}$. Then $\tilde{P}(x, w)$ is a simple Weierstrass polynomial over $X$, and by mapping $z$ to $w - \frac{a_1(x)}{n}$ we define an isomorphism of $C(X)$–algebras

$$C(X)[z]/(P(x, z)) \cong C(X)[w]/(\tilde{P}(x, w)) .$$

According to Corollary 3.5, the polynomial covering maps associated with $P(x, z)$ and $\tilde{P}(x, w)$ are therefore equivalent.

$\underline{B}$ (Direct proof). Define $\tilde{P}(x, z)$ as above. An explicit equivalence between the associated polynomial covering maps,

$$E \xrightarrow{\ h\ } \tilde{E}$$
$$\pi \searrow \swarrow \tilde{\pi}$$
$$X \quad,$$

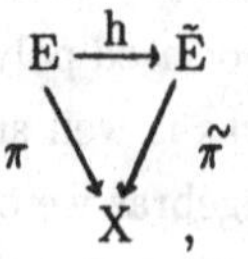

can then be defined by $h(x, z) = \left(x, z + \dfrac{a_1(x)}{n}\right)$ .

For the proof in B we do not need $X$ to be compact.

This proves Theorem 5.1.

In the next application, we shall give in Theorem 5.3 below a very explicit description of all 2–fold polynomial covering maps. But first a few preparations.

Let $S^1$ denote the unit circle considered as the space of complex numbers of modulus 1. For each $n \geq 1$ there is an n–fold covering map $p_n: S^1 \to S^1$ defined by $p_n(z) = z^n$ for $z \in S^1$ . The covering map $p_n$ is equivalent to the polynomial covering map $\pi_n: E_n \to S^1$ associated with the simple Weierstrass polynomial $P_n(x, w) = w^n - x$ , $x \in S^1$ , $w \in \mathbb{C}$ , over $S^1$ . A specific equivalence $h: S^1 \to E_n$ can be defined by $h(z) = (z^n, z)$ .

**Lemma 5.2.** Let $q: X \to \mathbb{C}^*$ be an arbitrary continuous map of the topological space $X$ into the nonzero complex numbers $\mathbb{C}^* = \mathbb{C}\backslash\{0\}$ . Then $P(x, z) = z^n - q(x)$ is a simple Weierstrass polynomial over $X$ , and the associated polynomial covering map $\pi: E \to X$ is equivalent to the pull–back of $p_n: S^1 \to S^1$ along the normalized map $\bar{q}: X \to S^1$ , where $\bar{q}(x) = \dfrac{q(x)}{|q(x)|}$ for all $x \in X$ .

Proof. It is clear that $P(x, z) = z^n - q(x)$ is a simple Weierstrass polynomial over $X$ . Define the map $f_q$ in the diagram

$$
\begin{array}{ccc}
E & \xrightarrow{\ f_q\ } & S^1 \\
\pi \downarrow & & \downarrow p_n \\
X & \xrightarrow[\ \bar{q}\ ]{} & S^1
\end{array}
$$

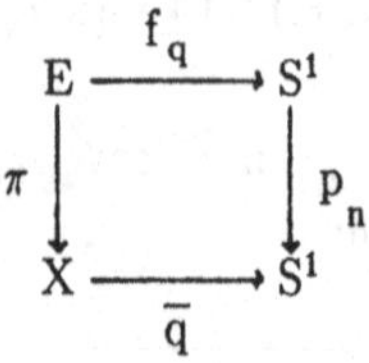

by $f_q(x, z) = \frac{z}{|z|}$ for $(x, z) \in E$.

The diagram is commutative, and $f_q$ maps fibres in $\pi$ bijectively onto fibres in $p_n$. Therefore it follows immediately that $\pi$ is equivalent to the pull–back of $p_n$ along $\bar{q}$ as asserted. This proves Lemma 5.2.

Let $P(x, z) = z^n + \sum_{i=1}^{n} a_i(x) z^{n-i}$ be a Weierstrass polynomial over the topological space $X$. As in §4 we get then the discriminant function

$$D = D(x) = \delta(a_1(x),\dots,a_n(x)) \colon X \to \mathbb{C} .$$

If $P(x, z)$ is simple, we remark, that $D$ will be a continuous map

$$D = D(x) \colon X \to \mathbb{C}^*$$

into the nonzero complex numbers $\mathbb{C}^* = \mathbb{C}\backslash\{0\}$.

Associated with the discriminant function we have the following Weierstrass polynomial over $X$,

$$P_{dr}(x, z) = z^n - D(x) ,$$

which we call <u>the discriminant radical</u> of the Weierstrass polynomial $P(x, z)$. If $P(x, z)$ is simple, $P_{dr}(x, z)$ is simple.

In some cases the discriminant radical gives complete topological information about the polynomial covering map associated with a simple Weierstrass polynomial. In particular we have

<u>Theorem 5.3.</u> Let $P(x, z) = z^2 + a_1(x) z + a_2(x)$ be an arbitrary simple Weierstrass polynomial of degree 2 over the (compact) space $X$. Then the associated polynomial covering map $\pi\colon E \to X$ is equivalent to the polynomial covering map $\pi_{dr}\colon E_{dr} \to X$ defined by the discriminant radical $P_{dr}(x, z)$ of $P(x, z)$.

More explicitely we have: The polynomial covering map $\pi\colon E \to X$ associated with the simple Weierstrass polynomial $P(x, z) = z^2 + a_1(x) z + a_2(x)$ is equivalent to the pull–back of $p_2\colon S^1 \to S^1$ along the normalized discriminant map $\bar{D}\colon X \to S^1$, where $D(x) = a_1(x)^2 - 4a_2(x)$ and $\bar{D}(x) = \frac{D(x)}{|D(x)|}$ for all $x \in X$.

<u>Proof.</u> <u>A</u> (Using characteristic algebras). By completion of the square, we can as usual rewrite $P(x, z)$ in the form

$$\begin{aligned} P(x, z) &= z^2 + a_1(x)\, z + a_2(x) \\ &= \left(z + \frac{a_1(x)}{2}\right)^2 - \frac{1}{4}\left(a_1(x)^2 - 4a_2(x)\right) \\ &= \left(z + \frac{a_1(x)}{2}\right)^2 - \frac{1}{4} D(x) , \end{aligned}$$

where $D(x) = a_1(x)^2 - 4a_2(x)$ is the discriminant function.

By mapping $z$ to $w - \frac{a_1(x)}{2}$ we can define an isomorphism of $C(X)$–algebras,

$$C(X)[z]/(P(x, z)) \cong C(X)[w]/(w^2 - \tfrac{D(x)}{4}) .$$

According to Corollary 3.5, the polynomial covering map $\pi\colon E \to X$ associated with $P(x, z)$ is therefore equivalent to the polynomial covering map associated with the simple Weierstrass polynomial $P'_{dr}(x, w) = w^2 - \frac{D(x)}{4}$ over $X$, which in turn by Lemma 5.2 is equivalent to the pull–back of $p_2\colon S^1 \to S^1$ along the normalized discriminant map $\overline{D}(x) = \frac{D(x)}{|D(x)|}\colon X \to S^1$. Also $\pi_{dr}\colon E_{dr} \to X$, which is associated with $P_{dr}(x, w) = w^2 - D(x)$ , is equivalent to this pull–back.

<u>B</u> (Direct proof). Rewrite $P(x, z)$ as above. It is then easy to see that an explicit equivalence between the polynomial covering maps associated with $P(x, z)$ and $P_{dr}(x, w)$,

$$\begin{array}{ccc} E & \xrightarrow{\;h\;} & E_{dr} \\ & {\scriptstyle\pi}\searrow \quad \swarrow{\scriptstyle\pi_{dr}} & \\ & X & \end{array} ,$$

can be defined by $h(x, z) = (x, 2z + a_1(x))$ .

For the proof in B we do not need $X$ to be compact.

This proves Theorem 5.3.

The result in Theorem 5.3 for simple Weierstrass polynomials of degree 2 invites you to pose the following

Problem ([35]): Determine the simple Weierstrass polynomials that define the same equivalence classes of polynomial covering maps as their descriminant radicals.

In relation to this problem we have the following result for Weierstrass polynomials of degree $n \geq 2$ .

Theorem 5.4. Let $P(x, z)$ be a simple Weierstrass polynomial over the (compact) space $X$ of the form $P(x, z) = z^n - q(x)$ , for an arbitrary nonzero continuous function $q\colon X \to \mathbb{C}^*$ and a degree $n \geq 2$ . Then the associated polynomial covering map $\pi\colon E \to X$ is equivalent to the polynomial covering map $\pi_{dr}\colon E_{dr} \to X$ defined by the discriminant radical $P_{dr}(x, z)$ of $P(x, z)$.

Before the proof of Theorem 5.4 we prove a general result about discriminants of ordinary complex polynomials.

Lemma 5.5. Let $P(z) = z^n + \sum_{i=1}^{n} a_i z^{n-i}$ be an ordinary complex polynomial of degree $n \geq 2$ , and let $\alpha_1,\dots,\alpha_n$ be the $n$ roots of $P(z)$ counted with multiplicity. Then the discriminant $D$ of $P(z)$ is given by

$$D = (-1)^{\frac{n(n-1)}{2}} \prod_{k=1}^{n} P'(\alpha_k) ,$$

where $P'(z)$ is the derivative of $P(z)$ .

If $P(z)$ has the special form $P(z) = z^n - q$ for a complex number $q$ , then the discriminant $D$ of $P(z)$ is given by

$$D = \begin{cases} n^n q^{n-1} & \text{for } n \equiv 1,2 \mod 4 \\ - n^n q^{n-1} & \text{for } n \equiv 0,3 \mod 4 . \end{cases}$$

Proof. For the general complex polynomial $P(z)$ we have the factorization

$$P(z) = \prod_{k=1}^{n} (z - \alpha_k) .$$

By the product formula for differentiation we get

$$P'(z) = \sum_{k=1}^{n} \Bigl( \prod_{\substack{l=1 \\ l \neq k}}^{n} (z - \alpha_l) \Bigr) .$$

Then $P'(\alpha_k) = \prod_{\substack{l=1 \\ l \neq k}}^{n} (\alpha_k - \alpha_l)$ for each $k = 1,\ldots,n$ .

Now a small computation:

$$\prod_{k=1}^{n} P'(\alpha_k) = \prod_{k=1}^{n} \Bigl( \prod_{\substack{l=1 \\ l \neq k}}^{n} (\alpha_k - \alpha_l) \Bigr) = \prod_{\substack{1 \leq k, l \leq n \\ l \neq k}} (\alpha_k - \alpha_l)$$

$$= (-1)^{\frac{n(n-1)}{2}} \prod_{1 \leq l < k \leq n} (\alpha_k - \alpha_l)^2 = (-1)^{\frac{n(n-1)}{2}} D .$$

Thereby we get the formula in the lemma.

Consider now a polynomial of the special form $P(z) = z^n - q$ for a complex number $q$ . Then $P'(z) = n\, z^{n-1}$ .

Choose a complex number $q_0$ such that $q_0^n = q$ , and let $w = e^{i\frac{2\pi}{n}}$ be a primitive n'th root of unity, i.e. $w^n = 1$ and $w^m \neq 1$ for $1 \leq m < n$ . Then $\alpha_k = q_0 w^k$ , $k = 1,\ldots,n$ , is the complete set of roots for $P(z) = z^n - q$ .

By the general formula for the discriminant and a small computation we get:

$$(-1)^{\frac{n(n-1)}{2}} D = \prod_{k=1}^{n} P'(\alpha_k) = \prod_{k=1}^{n} [n(q_0\, w^k)^{n-1}]$$

$$= n^n\, q_0^{n(n-1)} \prod_{k=1}^{n} (w^{-k}\, w^{kn})$$

$$= n^n\, q^{n-1} \left(\prod_{k=1}^{n} w^k\right)^{-1}$$

$$= n^n\, q^{n-1} \left(w^{\frac{n(n+1)}{2}}\right)^{-1} .$$

Now $w^{\frac{n(n+1)}{2}} = \left(e^{i\frac{2\pi}{n}}\right)^{\frac{n(n+1)}{2}} = e^{i\pi(n+1)} = (-1)^{n+1}$ .

Therefore

$$D = (-1)^{\frac{n(n-1)}{2}} (-1)^{n+1}\, n^n\, q^{n-1} = (-1)^{\frac{n^2}{2} + \frac{n}{2} + 1}\, n^n\, q^{n-1} .$$

Write $n = 4k + p$ with $k$ a nonnegative integer and $p = 0,1,2,3$. Then

$$(-1)^{\frac{n^2}{2} + \frac{n}{2} + 1} = (-1)^{\frac{p^2}{2} + \frac{p}{2} + 1} = \begin{cases} 1 & \text{for } p = 1,2 \\ -1 & \text{for } p = 0,3 \end{cases} .$$

Thereby we get the formula for $D$ in the special case stated in the lemma.

This proves Lemma 5.5.

Proof of Theorem 5.4. Let $P(x, z)$ be a simple Weierstrass polynomial over $X$ of the form $P(x, z) = z^n - q(x)$ for an arbitrary nonzero continuous function $q: X \to \mathbf{C}^*$. By Lemma 5.5, the discriminant function $D: X \to \mathbf{C}^*$ of $P(x, z)$ is then given by $D(x) = a \cdot q(x)^{n-1}$, where $a \in \mathbf{C}^*$ is a certain nonzero complex number. We shall prove that the polynomial covering map $\pi_{dr}: E_{dr} \to X$ defined by the discriminant radical $P_{dr}(x, z)$ of $P(x, z)$ is equivalent to the polynomial covering map $\pi: E \to X$ defined by $P(x, z)$. In view of Corollary III.3.3, it clearly suffices to prove that $\pi: E \to X$ is equivalent to the polynomial covering map $\pi'_{dr}: E'_{dr} \to X$ associated with the simple Weierstrass polynomial $P'_{dr}(x, w) = w^n - q(x)^{n-1}$ .

A (Using characteristic algebras). According to Corollary 3.5 we just have to produce an isomorphism of $C(X)$–algebras,

$$C(X)[z]/(z^n - q(x)) \underset{\psi}{\overset{\Phi}{\rightleftarrows}} C(X)[w]/(w^n - q(x)^{n-1}) \ ,$$

defined by a pair of inverse $C(X)$–algebra homomorphisms $\Phi$ and $\psi$.

A small computation shows first that we get well defined $C(X)$–algebra homomorphisms

$$\Phi \text{ defined by mapping } z \text{ to } \frac{w^{n-1}}{q(x)^{n-2}}$$

and $$\psi \text{ defined by mapping } w \text{ to } z^{n-1} \ .$$

A second small computation shows next that $\Phi$ and $\psi$ is indeed a pair of inverse $C(X)$–algebra homomorphisms.

B (Direct proof). An explicit equivalence of covering maps,

$$\begin{array}{ccc} E & \xrightarrow{h} & E'_{dr} \\ & {\scriptstyle\pi}\searrow \quad \swarrow{\scriptstyle\pi'_{dr}} & \\ & X & \end{array} ,$$

can be defined by $h(x, z) = (x, z^{n-1})$. The inverse map is given by $h^{-1}(x, w) = (x, q(x)\, w^{-1})$.

For the proof in B we do not need $X$ to be compact.

This proves Theorem 5.4.

## EXERCISES

1. Let $\pi: E \to X$ be a covering map onto a completely regular space $X$. Prove that if $e_1, e_2 \in E$ are points in $E$ such that $g(e_1) = g(e_2)$ for all $g \in C(E)$, then $e_1 = e_2$.

2. Prove the theorem of Gelfand and Kolmogoroff as outlined in §3.

3. Let $f(z) \in R[z]$ be a monic polynomial over the field $R$. Prove that the algebra $R[z]/(f(z))$ is separable in the classical sense if and only if $f(z)$ has no multiple roots in any field extension $K$ of $R$.

4. Prove that a topological space $E$ is connected if and only if the ring of continuous functions $C(E)$ has no other idempotents than $0$ and $1$.

5. ([35], Theorem 4.1). Prove that if a simple Weierstrass polynomial $P(x, z)$ over a topological space $X$ is defined by substitution of a Weierstrass polynomial $P_2(x, z)$ into a Weierstrass polynomial $P_1(x, z)$, i.e. $P(x, z) = P_1(x, P_2(x, z))$, then the associated polynomial covering map $\pi: E \to X$ decomposes into a corresponding composition $\pi = \pi_1 \circ \pi_2$ of polynomial covering maps.

6. Let $\pi_1: E_1 \to X$ and $\pi_2: E_2 \to E_1$ be polynomial covering maps. Prove that the composition $\pi = \pi_1 \circ \pi_2: E_2 \to X$ is a polynomial covering map. (The converse is also true, [41].)

7. Let $P(x, z)$ be an irreducible, simple Weierstrass polynomial over a compact, connected space $X$, and assume that the associated polynomial covering map $\pi: E \to X$ is a regular covering map, i.e. $\pi_*(\pi_1(E))$ is a normal subgroup in $\pi_1(X)$. Define the simple Weierstrass polynomial $\overline{P}(e, z)$ over $E$ by $\overline{P}(e, z) = P(\pi(e), z)$ for $e \in E$ and $z \in \mathbb{C}$. Prove that $\overline{P}(e, z)$ is completely solvable on $E$.

8. ([32], §4). Let $B^n = \mathbb{C}^n \backslash \Delta$ denote the complement of the discriminant set $\Delta$ in complex n–space $\mathbb{C}^n$, and let $\mathbb{C}^* = \mathbb{C} \backslash \{0\}$ denote the nonzero complex numbers. Define the map $\delta: B^n \to \mathbb{C}^*$ by associating to $(a_1,\ldots,a_n) \in B^n$ the discriminant

$\delta(a_1,...,a_n) \in \mathbf{C}^*$ of the polynomial $P(z) = z^n + \sum_{i=1}^{n} a_i z^{n-1}$. Let $B_1^n = \delta^{-1}(1)$ be the inverse image of $1 \in \mathbf{C}^*$.

(i) Prove that

$$\delta(w\, a_1, w^2\, a_2,...,w^n\, a_n) = w^{n(n-1)}\, \delta(a_1, a_2,...,a_n)$$

for all $w \in \mathbf{C}^*$ and $(a_1, a_2,...,a_n) \in B^n$.

(ii) Prove that $B_1^n$ is pathwise connected.

(iii) Prove that $\delta\colon B^n \to \mathbf{C}^*$ is a locally trivial fibration.

(iv) Prove that $B_1^n$ is an Eilenberg – MacLane space of type $([B(n), B(n)], 1)$, where $[B(n), B(n)]$ denotes the commutator subgroup in $B(n)$.

# APPENDIX 1

## A presentation of the abstract coloured braid group

by Lars Gæde

The purpose of this appendix is to derive the presentation stated in Lemma I.4.2, p. 20, of the abstract coloured braid group, $H_n$, which geometrically corresponds to those braids that leave the order of the strings unchanged. There appears to be slight mistakes in several of the presentations found in the literature, and hence it seems appropriate to give a detailed derivation, so as to make it possible for the reader to check our computations.

Our main tool will be a general method of finding presentations of subgroups, called the Reidemeister – Schreier Rewriting Process. We apply this – not to $H_n$ directly, since the computations involved are rather messy – but instead to the group $D_n$ of braids that, geometrically speaking, do not change the position of the n'th string. The presentation obtained shows that $D_n$ is isomorphic to a semi–direct product of the abstract braid group on $n-1$ strings and a free group, which in turn implies that $H_n$ is isomorphic to a semi–direct product of $H_{n-1}$ and a free group. Using this as the basis of an induction argument finally yields the desired presentation of $H_n$.

References for the material in this appendix are:


W.L. Chow, "On the algebraic braid group", Ann. of Math. 49(1948), 654–658.

D.L. Johnson, "Topics in the theory of group presentations", Cambridge University Press, 1980.

H. Zieschang, E. Vogt, H.–D. Coldewey, "Surfaces and planar discontinuous groups", Springer–Verlag, 1980.


### 1. The Reidemeister – Schreier Rewriting Process

We begin with a description of the Reidemeister – Schreier Rewriting Process, which is a general method of finding a presentation of a subgroup of a group for which a presentation is known. An alternative, geometric, description can be found in the book by Zieschang et al.

Let us first consider the case of a subgroup $L$ of $F(X)$, the free group with the set $X$ as generators. A (right) *Schreier transversal* of $L$ is a subset $\mathcal{T} \subseteq F(X)$, consisting of exactly one representative from each (right) coset of $L$, and having the property that if $T \in \mathcal{T}$ and $T$ can be written as a *reduced* word $x_1^{\varepsilon_1} \dots x_n^{\varepsilon_n}$ with $x_i \in X$, $\varepsilon_i = \pm 1$, then also $1, x_1^{\varepsilon_1}, x_1^{\varepsilon_1} x_2^{\varepsilon_2}, \dots, x_1^{\varepsilon_1} \dots x_{n-1}^{\varepsilon_{n-1}} \in \mathcal{T}$. It is not hard to show that a Schreier transversal can always be found.

For any $U \in F(X)$ we let $\overline{U} \in \mathcal{T}$ denote the representative of the coset $LU$. Now, define the set

$$Y = \{(T, x) \in \mathcal{T} \times X \mid \overline{Tx} \neq Tx\}$$

and a homomorphism $\phi: F(Y) \to L$ by

$$\phi(T, x) = Tx(\overline{Tx})^{-1} .$$

$\phi$ really does map into $L \subseteq F(X)$, since

$$L\phi(T, x) = L(Tx(\overline{Tx})^{-1}) = (L(Tx))(\overline{Tx})^{-1} = (L(\overline{Tx}))(\overline{Tx})^{-1} = L(\overline{Tx}(\overline{Tx})^{-1}) = L .$$

Given a $U \in L$ we write $U$ as a reduced word $x_1^{\varepsilon_1} \dots x_n^{\varepsilon_n}$ and set up the following table

| $T_1$ | $T_2$ | ... | $T_n$ | $T_{n+1}$ |
|---|---|---|---|---|
| $x_1^{\varepsilon_1}$ | $x_2^{\varepsilon_2}$ | ... | $x_n^{\varepsilon_n}$ | |
| $V_1$ | $V_2$ | ... | $V_n$ | |

The entries in the top row are computed in the following way:

$$T_1 = 1 \ , \ T_{i+1} = \overline{T_i \ x_i^{\varepsilon_i}} \ , 1 \le i \le n \ .$$

Notice, that since

$$LT_{i+1} = L(\overline{T_i \ x_i^{\varepsilon_i}}) = L(T_i \, x_i^{\varepsilon_i}) = (LT_i) \, x_i^{\varepsilon_i} = \dots = (LT_1) \, x_1^{\varepsilon_1} \dots x_i^{\varepsilon_i} = L(x_1^{\varepsilon_1} \dots x_i^{\varepsilon_i}) \ ,$$

we have

$$T_{i+1} = \overline{x_1^{\varepsilon_1} \ldots x_i^{\varepsilon_i}} ,$$

and in particular

$$T_{n+1} = \overline{x_1^{\varepsilon_1} \ldots x_n^{\varepsilon_n}} = \overline{U} = 1 ,$$

as $U \in L$.

The bottom row is now computed as follows:

Let

$$\binom{T}{x} = \begin{cases} (T, x) & \text{if } \ \overline{Tx} \neq Tx \\ 1 & \text{otherwise} \end{cases} ,$$

and

$$V_i = \begin{cases} \binom{T_i}{x_i} & \text{if } \varepsilon_i = +1 \\ \binom{T_{i+1}}{x_i}^{-1} & \text{if } \varepsilon_i = -1 \end{cases} , 1 \leq i \leq n .$$

Now, if $\varepsilon_i = +1$ we have

$$\phi V_i = T_i\, x_i (\overline{T_i\ x_i})^{-1} = T_i\, x_i^{\varepsilon_i} (\overline{T_i\ x_i^{\varepsilon_i}})^{-1} = T_i\, x_i^{\varepsilon_i}\, T_{i+1}^{-1} ,$$

and if $\varepsilon_i = -1$ we again have

$$\phi V_i = \overline{T_{i+1}\ x_i}\, x_i^{-1}\, T_{i+1}^{-1} = \overline{T_{i+1}\ x_i^{-\varepsilon_i}}\, x_i^{\varepsilon_i}\, T_{i+1}^{-1} = T_i\, x_i^{\varepsilon_i}\, T_{i+1}^{-1} ,$$

since

$$L(T_{i+1}\, x_i^{-\varepsilon_i}) = (LT_{i+1})\, x_i^{-\varepsilon_i} = LT_i .$$

Therefore, if we let $V = V_1 \ldots V_n$, we have

$$\phi V = (T_1 x_1^{\varepsilon_1} T_2^{-1})(T_2 x_2^{\varepsilon_2} T_3^{-1}) \ldots (T_n x_n^{\varepsilon_n} T_{n+1}^{-1}) = T_1\, x_1^{\varepsilon_1} \ldots x_n^{\varepsilon_n}\, T_{n+1}^{-1} = 1 x_1^{\varepsilon_1} \ldots x_n^{\varepsilon_n}\, 1 = U .$$

So we define the "rewriting map" $\psi: L \to F(Y)$ by $\psi U = V$ , and we have just seen that $\phi \circ \psi = id_L$ .

In the definition of $\psi$ we assumed that $x_1^{\varepsilon_1}...x_n^{\varepsilon_n}$ was a reduced word. This is not necessary, for consider the words $x_1^{\varepsilon_1}...x_n^{\varepsilon_n} x^{\varepsilon} x^{-\varepsilon} \tilde{x}_1^{\omega_1}...\tilde{x}_m^{\omega_m}$ and $x_1^{\varepsilon_1}...x_n^{\varepsilon_n} \tilde{x}_1^{\omega_1}...\tilde{x}_m^{\omega_m}$ . If the table constructed from the last word is

| $T_1$ | ... | $T_n$ | $\tilde{T}_1$ | ... | $\tilde{T}_m$ | $\tilde{T}_{m+1}$ |
|---|---|---|---|---|---|---|
| $x_1^{\varepsilon_1}$ | ... | $x_n^{\varepsilon_n}$ | $\tilde{x}_1^{\omega_1}$ | ... | $\tilde{x}_m^{\omega_m}$ | |
| $V_1$ | ... | $V_n$ | $\tilde{V}_1$ | ... | $\tilde{V}_m$ | |

then the table constructed from the first word will be

| $T_1$ | ... | $T_n$ | $\tilde{T}_1$ | $T$ | $\tilde{T}_1$ | ... | $\tilde{T}_m$ | $\tilde{T}_{m+1}$ |
|---|---|---|---|---|---|---|---|---|
| $x_1^{\varepsilon_1}$ | ... | $x_n^{\varepsilon_n}$ | $x^{\varepsilon}$ | $x^{-\varepsilon}$ | $\tilde{x}_1^{\omega_1}$ | ... | $\tilde{x}_m^{\omega_m}$ | |
| $V_1$ | ... | $V_n$ | $V'$ | $V''$ | $\tilde{V}_1$ | ... | $\tilde{V}_m$ | |

where

$$T = \overline{x_1^{\varepsilon_1} \dots x_n^{\varepsilon_n} x^{\varepsilon}} \ ,$$

$$V' = \begin{cases} \begin{pmatrix} \tilde{T}_1 \\ x \end{pmatrix} & \text{if } \varepsilon = +1 \\ \\ \begin{pmatrix} T \\ x \end{pmatrix}^{-1} & \text{if } \varepsilon = -1 \end{cases} \ , \quad V'' = \begin{cases} \begin{pmatrix} T \\ x \end{pmatrix} & \text{if } -\varepsilon = +1 \\ \\ \begin{pmatrix} \tilde{T}_1 \\ x \end{pmatrix}^{-1} & \text{if } -\varepsilon = -1 \end{cases} \ .$$

So $V'' = V'^{-1}$ , and hence $V_1...V_n V' V'' \tilde{V}_1...\tilde{V}_m = V_1...V_n \tilde{V}_1...\tilde{V}_m$ .

We now want to show that $\psi \circ \phi = id_{F(Y)}$ . Since it is easily shown, by constructing the relevant tables, that $\psi$ is a homomorphism, it suffices to consider a generator $y = (T,x) \in Y$ . Then $\phi y = Tx(\overline{Tx})^{-1}$ . Now, let $T = x_1^{\varepsilon_1}...x_n^{\varepsilon_n}$ , $\overline{Tx} = \tilde{x}_1^{\omega_1}...\tilde{x}_m^{\omega_m}$ be reduced words. Then because of the Schreier property of $\mathcal{T}$ we get the following table for $\phi y$ :

| $1$ | $x_1^{\varepsilon_1}$ | $x_1^{\varepsilon_1} x_2^{\varepsilon_2}$ | ... | $x_1^{\varepsilon_1} \ldots x_{n-1}^{\varepsilon_{n-1}}$ | $T$ | $\tilde{x}_1^{\omega_1} \ldots \tilde{x}_m^{\omega_m}$ | $\tilde{x}_1^{\omega_1} \ldots \tilde{x}_{m-1}^{\omega_{m-1}}$ | ... | $\tilde{x}_1^{\omega_1}$ | $1$ |
|---|---|---|---|---|---|---|---|---|---|---|
| $x_1^{\varepsilon_1}$ | $x_2^{\varepsilon_2}$ | $x_3^{\varepsilon_3}$ | ... | $x_n^{\varepsilon_n}$ | $x$ | $\tilde{x}_m^{-\omega_m}$ | $\tilde{x}_{m-1}^{-\omega_{m-1}}$ | ... | $\tilde{x}_1^{-\omega_1}$ | |
| $1$ | $1$ | $1$ | ... | $1$ | $y$ | $1$ | $1$ | ... | $1$ | |

whereby $\psi\phi y = y$ .

This proves the

<u>Nielsen–Schreier theorem.</u> *There is an isomorphism of groups* $L \cong F(Y)$ .

We now turn to the general case of a subgroup $H$ of a group $G$ with presentation $\langle X \mid \mathcal{R} \rangle$ . This means that we have an exact sequence

$$1 \to \langle \mathcal{R}^{F(X)} \rangle \xrightarrow{\subseteq} F(X) \xrightarrow{\lambda} G \to 1 \ .$$

$L = \lambda^{-1}\{H\}$ is a subgroup of $F(X)$ , and the preceding theorem gives us an isomorphism $\phi\colon F(Y) \to L$ . Setting $\mu = \lambda \circ \phi\colon F(Y) \to H$ , we then have the exact sequence

$$1 \to \operatorname{Ker} \mu \xrightarrow{\subseteq} F(Y) \xrightarrow{\mu} H \to 1 \ ,$$

and all that remains is to find a set of relators $\mathcal{Q} \in F(Y)$ , such that $\operatorname{Ker} \mu = \langle \mathcal{Q}^{F(Y)} \rangle$ . So, let $V \in \operatorname{Ker} \mu$ . Then $\phi V \in \operatorname{Ker} \lambda = \langle \mathcal{R}^{F(X)} \rangle$ , and hence

$$\phi V = (U_1 R_1^{\varepsilon_1} U_1^{-1}) \ldots (U_n R_n^{\varepsilon_n} U_n^{-1})$$

for some $U_i \in F(X)$ , $R_i \in \mathcal{R}$ , $\varepsilon_i = \pm 1$ . In order to be able to use $\phi^{-1}$ , we write $U_i$ as $\tilde{U}_i T_i$ with $\tilde{U}_i \in L$ , $T_i \in \mathcal{T}$, so that

$$\phi V = \tilde{U}_1 (T_1 R_1 T_1^{-1})^{\varepsilon_1} \tilde{U}_1^{-1} \ldots \tilde{U}_n (T_n R_n T_n^{-1})^{\varepsilon_n} \tilde{U}_n^{-1} \ ,$$

and since now both $\tilde{U}_i \in L$ and $T_i R_i T_i^{-1} \in \operatorname{Ker} \lambda \subseteq L$ , we get, with $V_i = \phi^{-1} \tilde{U}_i$ ,

$$V = V_1 \phi^{-1}(T_1 R_1 T_1^{-1})^{\varepsilon_1} V_1^{-1} \dots V_n \phi^{-1}(T_n R_n T_n^{-1})^{\varepsilon_n} V_n^{-1} \ .$$

This shows that if we put

$$\mathcal{Q} = \{\phi^{-1}(TRT^{-1}) \mid T \in \mathcal{T}, R \in \mathcal{R}\} \ ,$$

we have $\operatorname{Ker} \mu \subseteq \langle \mathcal{Q}^{F(Y)} \rangle$ . And the reverse inclusion is obvious, for if $Q \in \mathcal{Q}$ we have $Q = \phi^{-1}(TRT^{-1})$ for some $T \in \mathcal{T}, R \in \mathcal{R}$ , and then

$$\mu Q = \lambda(TRT^{-1}) = e \ .$$

So $\operatorname{Ker} \mu$ is a normal subgroup of $F(Y)$ containing the set $\mathcal{Q}$, and $\langle \mathcal{Q}^{F(Y)} \rangle$ is the smallest such subgroup.

This proves the

Reidemeister–Schreier theorem. *The subgroup* $H$ *can be presented as* $\langle Y \mid \mathcal{Q} \rangle$ .

Sometimes it is possible to simplify this presentation by making use of the relators in $\mathcal{R}$. We therefore imagine ourselves given a set $Z$ , an epimorphism $\chi: F(Y) \to F(Z)$ , and a homomorphism $\nu: F(Z) \to H$ , such that the diagram

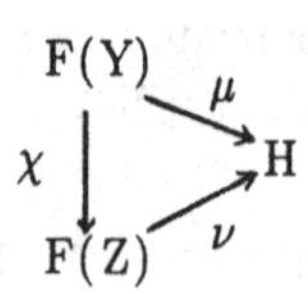


commutes, i.e. $\nu \circ \chi = \mu = \lambda \circ \phi$ . We see that $\nu$ is surjective, so that we have an exact sequence

$$1 \longrightarrow \operatorname{Ker} \nu \xrightarrow{\subseteq} F(Z) \xrightarrow{\nu} H \longrightarrow 1 \ .$$

Let $W \in \operatorname{Ker} \nu$ . Since $\chi$ is surjective, $W = \chi V$ for some $V \in F(Y)$ . As

$$e = \nu W = \mu V \ ,$$

we have $V \in \operatorname{Ker} \mu = \langle \mathcal{Q}^{F(Y)} \rangle$ , so

$$V = (V_1 Q_1^{\varepsilon_1} V_1^{-1})...(V_n Q_n^{\varepsilon_n} V_n^{-1})$$

with $V_i \in F(Y)$, $Q_i \in \mathcal{Q}$, $\varepsilon_i = \pm 1$. Letting $W_i = \chi V_i$ we get

$$W = (W_1 (\chi Q_1)^{\varepsilon_1} W_1^{-1})...(W_n (\chi Q_n)^{\varepsilon_n} W_n^{-1}) ,$$

so if we put

$$\mathcal{P} = \chi\{\mathcal{Q}\}\backslash 1 ,$$

we have $\operatorname{Ker} \nu \subseteq \langle \mathcal{P}^{F(Z)} \rangle$, and the reverse inclusion follows as before. We then have the

Proposition. H *can be presented as* $\langle Z | \mathcal{P} \rangle$ .

## 2. A presentation of $D_n$.

Recall that the abstract braid group $B_n$ is given by the presentation $\langle \sigma_1,...,\sigma_{n-1} | \mathcal{R}_1 \cup \mathcal{R}_2 \rangle$, where

$$\mathcal{R}_1 = \{\sigma_j \sigma_k \sigma_j^{-1} \sigma_k^{-1} | 1 \leq j, k \leq n-1, k > j+1\} ,$$
$$\mathcal{R}_2 = \{ \sigma_j \sigma_{j+1} \sigma_j \sigma_{j+1}^{-1} \sigma_j^{-1} \sigma_{j+1}^{-1} | 1 \leq j \leq n-2\} ,$$

i.e. we have an exact sequence

$$1 \rightarrow \langle (\mathcal{R}_1 \cup \mathcal{R}_2)^{F(\sigma_1,...,\sigma_{n-1})} \rangle \xrightarrow{\subseteq} F(\sigma_1,...,\sigma_{n-1}) \xrightarrow{\lambda} B_n \rightarrow 1 .$$

Also, there is the permutation homomorphism $\tau: B_n \rightarrow \Sigma_n$, given by

$$\tau(\lambda\sigma_i) = (i\ i{+}1) ,$$

which is used to define the abstract coloured braid group $H_n = \operatorname{Ker} \tau$. Now, in principle the Reidemeister–Schreier rewriting process could be applied directly to $H_n$, but as the computations are rather difficult to handle, we will use a different approach due to Chow.

Inside $\Sigma_n$ we have a subgroup, which in an obvious manner can be identified with $\Sigma_{n-1}$, namely those permutations, where $n \mapsto n$ . We use this to define the subgroup

$$D_n = \tau^{-1}\{\Sigma_{n-1}\} \ .$$

Geometrically, $D_n$ consists of those braids that do not change the position of the n'th string.

To find a presentation of $D_n$ we first of all need a Schreier transversal of $\lambda^{-1}\{D_n\}$ . We take $\mathcal{S} = \{S_0,\dots,S_{n-1}\}$ where

$$S_i = \sigma_{n-1}\cdots\sigma_{n-i} \ , \quad 0 \le i \le n-1 \ .$$

Of course, the right hand side doesn't really make sense when $i = 0$ , and should in this case be interpreted as a 1 . This convention will also be used in the sequel.

Since

$$\tau\lambda S_i = (n-1\ n)(n-2\ n-1)\dots(n-i\ n-i+1) = (n-i\ n-i+1\dots n-1\ n) \ ,$$

we see that the different $S_i$'s must belong to different $\lambda^{-1}\{D_n\}$ – cosets, as $\tau \circ \lambda$ takes different $S_i$'s to different $\Sigma_{n-1}$ – cosets. And since $\tau \circ \lambda$ is an epimorphism we have $|F(\sigma_1,\dots,\sigma_{n-1})\colon \lambda^{-1}\{D_n\}| = |\Sigma_n\colon \Sigma_{n-1}| = n$ , so that there is an element of $\mathcal{S}$ in each $\lambda^{-1}\{D_n\}$ – coset. Finally, $\mathcal{S}$ clearly has the Schreier property.

We must now calculate $\overline{S_i\sigma_j^\varepsilon}$ for all $0 \le i \le n-1$ , $1 \le j \le n-1$ , $\varepsilon = \pm 1$ . As the permutation $\tau\lambda(S_i\,\sigma_j^\varepsilon)$ acts on n as follows

$$n \xrightarrow{\tau\lambda S_i} n-i \xrightarrow{\tau\lambda\sigma_j^\varepsilon} \begin{cases} n-i & \text{if } n-i \ne j,\ j+1 \\ n-i+1 & \text{if } n-i = j \\ n-i-1 & \text{if } n-i = j+1 \end{cases} \ ,$$

we must have

$$\overline{S_i\sigma_j^\varepsilon} = \begin{cases} S_i & \text{if } i+j \ne n-1,\ n \\ S_{i+1} & \text{if } i+j = n-1 \\ S_{i-1} & \text{if } i+j = n \end{cases} \ .$$

Now we can go on to calculate $S_i\sigma_j(\overline{S_i\sigma_j})^{-1}$ . If $i + j \neq n - 1, n$ , we get

$$S_i \sigma_j S_i^{-1} = (\sigma_{n-1} \cdots \sigma_{n-i}) \sigma_j (\sigma_{n-i}^{-1} \cdots \sigma_{n-1}^{-1}) \ ;$$

if $i+j = n-1$ , we get

$$S_{n-j-1} \sigma_j S_{n-j}^{-1} = (\sigma_{n-1} \cdots \sigma_{j+1}) \sigma_j (\sigma_j^{-1} \ldots \sigma_{n-1}^{-1}) = 1 \ ;$$

and finally if $i+j = n$ , we get

$$S_{n-j} \sigma_j S_{n-j-1}^{-1} = (\sigma_{n-1} \cdots \sigma_j) \sigma_j (\sigma_{j+1}^{-1} \cdots \sigma_{n-1}^{-1}) \ .$$

Following the procedure described in the previous section, we should therefore put

$$Y = \{(S_i, \sigma_j) \mid 0 \leq i \leq n-1 \ , \ 1 \leq j \leq n-1 \ , \ i+j \neq n-1\}$$

and define $\phi$: $F(Y) \to \lambda^{-1}\{D_n\}$ by

$$\phi(S_i, \sigma_j) = \begin{cases} \sigma_{n-1} \cdots \sigma_{n-i} \ \sigma_j \ \sigma_{n-i}^{-1} \cdots \sigma_{n-1}^{-1} & \text{if } i+j \neq n \\ \sigma_{n-1} \cdots \sigma_{j+1} \ \sigma_j^2 \ \sigma_{j+1}^{-1} \cdots \sigma_{n-1}^{-1} & \text{if } i+j = n \end{cases} .$$

At this point we notice that if $i+j < n-1$ , then $n-i > j+1$ and hence

$$\begin{aligned} \lambda(\sigma_{n-1} \cdots \sigma_{n-i} \ \sigma_j \ \sigma_{n-i}^{-1} \cdots \sigma_{n-1}^{-1}) &= \lambda(\sigma_{n-1} \cdots \sigma_j \ \sigma_{n-i} \ \sigma_{n-i}^{-1} \cdots \sigma_{n-1}^{-1}) \\ &= \cdots \\ &= \lambda \ \sigma_j \end{aligned}$$

by repeated use of the relations in $\mathscr{R}_1$ . Similarly, if $i+j > n$ , then $j > n-i$ , and

$$\begin{aligned} \lambda(\sigma_{n-1} \cdots \sigma_{n-i} \ \sigma_j \ \sigma_{n-i}^{-1} \cdots \sigma_{n-1}^{-1}) &\overset{1}{=} \lambda(\sigma_{n-1} \cdots \sigma_{j-1} \ \sigma_j \ \sigma_{j-1}^{-1} \cdots \sigma_{n-1}^{-1}) \\ &= \lambda(\sigma_{n-1} \cdots \sigma_{j+1} \ \sigma_j \ \sigma_{j-1} \ \sigma_j \ \sigma_{j-1}^{-1} \ \sigma_j^{-1} \ \sigma_{j+1}^{-1} \cdots \sigma_{n-1}^{-1}) \\ &\overset{2}{=} \lambda(\sigma_{n-1} \cdots \sigma_{j+1} \ \sigma_{j-1} \ \sigma_j \ \sigma_{j-1} \ \sigma_{j-1}^{-1} \ \sigma_j^{-1} \ \sigma_{j+1}^{-1} \cdots \sigma_{n-1}^{-1}) \\ &= \lambda(\sigma_{n-1} \cdots \sigma_{j+1} \ \sigma_{j-1} \ \sigma_{j+1}^{-1} \cdots \sigma_{n-1}^{-1}) \\ &\overset{1}{=} \lambda \ \sigma_{j-1} \ , \end{aligned}$$

where 1 follows as above, and 2 follows by the relations in $\mathcal{R}_2$ .

We will therefore use the proposition of the previous section to simplify the presentation. So we let

$$Z = \{\sigma_1,\dots,\sigma_{n-2} , a_1,\dots,a_{n-1}\}$$

and define a homomorphism $\chi$: F(Y) $\to$ F(Z) by

$$\chi(S_i, \sigma_j) = \begin{cases} \sigma_j & \text{if } i+j < n-1 \\ a_j & \text{if } i+j = n \\ \sigma_{j-1} & \text{if } i+j > n \end{cases} ,$$

and a homomorphism $\nu$: F(Z) $\to$ $D_n$ by

$$\nu\sigma_j = \lambda\sigma_j \ , \ \nu a_j = \lambda(\sigma_{n-1}\dots\sigma_{j+1}\,\sigma_j^2\,\sigma_{j+1}^{-1}\dots\sigma_{n-1}^{-1}) \ .$$

Clearly, $\chi$ is surjective, and $\nu \circ \chi = \lambda \circ \phi$ .

The final step is to compute the relators

$$\mathcal{P} = \{\chi\phi^{-1}(TRT^{-1})\,|\,T \in \mathcal{T}, R \in \mathcal{R}_1 \cup \mathcal{R}_2\}\backslash 1 \ .$$

$\phi^{-1}$ is computed by means of tables as shown in the last section, and we note that

$$\chi\begin{pmatrix} S_i \\ \sigma_j \end{pmatrix} = \begin{cases} \sigma_j & \text{if } i+j < n-1 \\ 1 & \text{if } i+j = n-1 \\ a_j & \text{if } i+j = n \\ \sigma_{j-1} & \text{if } i+j > n \end{cases} .$$

The results are as follows:

| | | $\chi\phi^{-1}(S_i\ \sigma_j\ \sigma_k\ \sigma_j^{-1}\ \sigma_k^{-1}\ S_i^{-1})$ , $0 \le i \le n-1$ , $1 \le j , k \le n-1$ , $k > j+1$ | |
|---|---|---|---|
| $i+j<n-1$ | $i+k<n-1$ | $\sigma_j\ \sigma_k\ \sigma_j^{-1}\ \sigma_k^{-1}$ , $1 \le j , k \le n-2$ , $k > j+1$ | (1) |
| | $i+k=n-1$ | 1 | (2) |
| | $i+k=n$ | $\sigma_j\ a_k\ \sigma_j^{-1}\ a_k^{-1}$ , $1 \le j \le n-2$ , $1 \le k \le n-1$ , $k > j+1$ | (3) |
| | $i+k>n$ | $\sigma_j\ \sigma_{k-1}\ \sigma_j^{-1}\ \sigma_{k-1}^{-1}$ , $1 \le j \le n-2$ , $2 \le k \le n-1$ , $k > j+2$ | (4) |
| $i+j=n-1$ | | 1 | (5) |
| $i+j=n$ | | $a_j\ \sigma_{k-1}\ a_j^{-1}\ \sigma_{k-1}^{-1}$ , $1 \le j \le n-1$ , $2 \le k \le n-1$ , $k > j+1$ | (6) |
| $i+j>n$ | | $\sigma_{j-1}\ \sigma_{k-1}\ \sigma_{j-1}^{-1}\ \sigma_{k-1}^{-1}$ , $2 \le j , k \le n-1$ , $k > j+1$ | (7) |

| | | $\chi\phi^{-1}(S_i\ \sigma_j\ \sigma_{j+1}\ \sigma_j\ \sigma_{j+1}^{-1}\ \sigma_j^{-1}\ \sigma_{j+1}^{-1}\ S_i^{-1})$ , $0 \le i \le n-1$ , $1 \le j \le n-2$ | |
|---|---|---|---|
| $i+j<n-1$ | $i+j<n-2$ | $\sigma_j\ \sigma_{j+1}\ \sigma_j\ \sigma_{j+1}^{-1}\ \sigma_j^{-1}\ \sigma_{j+1}^{-1}$ , $1 \le j \le n-2$ | (8) |
| | $i+j=n-2$ | 1 | (9) |
| $i+j=n-1$ | | $\sigma_j\ a_j\ \sigma_j^{-1}\ a_{j+1}^{-1}$ , $1 \le j \le n-2$ | (10) |
| $i+j=n$ | | $a_j\ a_{j+1}\ \sigma_j\ a_{j+1}^{-1}\ a_j^{-1}\ \sigma_j^{-1}$ , $1 \le j \le n-2$ | (11) |
| $i+j>n$ | | $\sigma_{j-1}\ \sigma_j\ \sigma_{j-1}\ \sigma_j^{-1}\ \sigma_{j-1}\ \sigma_j^{-1}$ , $2 \le j \le n-1$ | (12) |

We illustrate the computations leading to (3). The table giving $\phi^{-1}(S_i\ \sigma_j\ \sigma_k\ \sigma_j^{-1}\ \sigma_k^{-1}\ S_i^{-1})$ looks like this:

| $S_0$ | $\cdots$ | $S_{i-1}$ | $S_i$ | $S_i$ | $S_{i-1}$ | $S_{i-1}$ | $S_i$ | $\cdots$ | $S_1$ | $S_0$ |
|---|---|---|---|---|---|---|---|---|---|---|
| $\sigma_{n-1}$ | $\cdots$ | $\sigma_{n-i}$ | $\sigma_j$ | $\sigma_k$ | $\sigma_j^{-1}$ | $\sigma_k^{-1}$ | $\sigma_{n-i}^{-1}$ | $\cdots$ | $\sigma_{n-1}^{-1}$ | |

$$\underbrace{\qquad}_{S_i} \qquad\qquad \underbrace{\qquad}_{S_i^{-1}}$$

$$\left(\begin{smallmatrix}S_0\\\sigma_{n-1}\end{smallmatrix}\right)\ \cdots\ \left(\begin{smallmatrix}S_{i-1}\\\sigma_{n-i}\end{smallmatrix}\right)\left(\begin{smallmatrix}S_i\\\sigma_j\end{smallmatrix}\right)\left(\begin{smallmatrix}S_i\\\sigma_k\end{smallmatrix}\right)\left(\begin{smallmatrix}S_{i-1}\\\sigma_j\end{smallmatrix}\right)^{-1}\left(\begin{smallmatrix}S_i\\\sigma_k\end{smallmatrix}\right)^{-1}\left(\begin{smallmatrix}S_{i-1}\\\sigma_{n-i}\end{smallmatrix}\right)^{-1}\ \cdots\ \left(\begin{smallmatrix}S_0\\\sigma_{n-1}\end{smallmatrix}\right)^{-1}$$

and using $\chi$ on the last row gives us

| 1 | $\cdots$ | 1 | $\sigma_j$ | $a_k$ | $\sigma_j^{-1}$ | $a_k^{-1}$ | 1 | $\cdots$ | 1 |
|---|---|---|---|---|---|---|---|---|---|

We have thus obtained the relator $\sigma_j\ a_k\ \sigma_j^{-1}\ a_k^{-1}$ for $i+j < n-1$ , $i+k=n$ , i.e. $k > j+1$ . The remaining calculations are left to the energetic reader. In connection with this we remark, that the tables will always have the same general appearance, namely

| 1 | | T | | T | | 1 |
|---|---|---|---|---|---|---|
| T | | R | | $T^{-1}$ | | |
| 1 | $\cdots$ | 1 | | 1 | $\cdots$ | 1 |

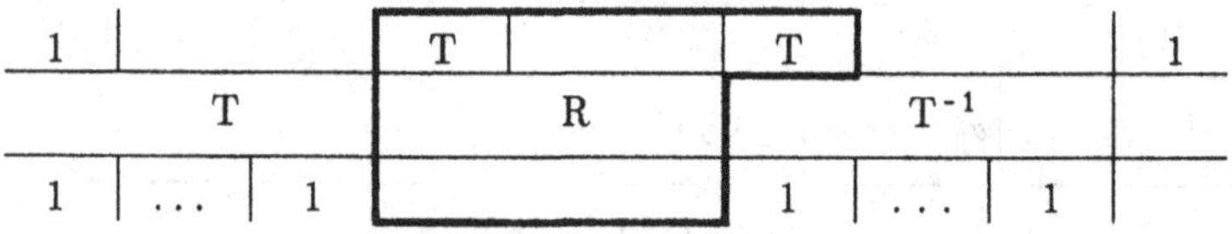

so that we need only write down the "framed" part.

Exercise: Why does the table corresponding to a word of the form $TRT^{-1}$ , $T \in \mathcal{T}$ , $R \in \mathcal{R}$ , look as claimed above?

Collecting the relators (1) – (12) we find

Theorem (Chow). $D_n$ *has a presentation with*

*generators:* $\sigma_1,\ldots,\sigma_{n-2}$ , $a_1,\ldots,a_{n-1}$

*relations:*

$$\sigma_j\,\sigma_k = \sigma_k\,\sigma_j \ , \ k > j+1 \ , \qquad (1), (4), (7)$$

$$\sigma_j\,\sigma_{j+1}\,\sigma_j = \sigma_{j+1}\,\sigma_j\,\sigma_{j+1} \ , \qquad (8), (12)$$

$$\sigma_j^{-1}\,a_k\,\sigma_j = a_k \ , \ k \neq j\,,\,j+1 \ , \qquad (3), (6)$$

$$\sigma_j^{-1}\,a_j\,\sigma_j = a_j\,a_{j+1}\,a_j^{-1} \ , \qquad (10)$$

$$\sigma_j^{-1}\,a_{j+1}\,\sigma_j = a_j \ , \qquad (11)$$

In order to proceed we now need to know something about presentations of semi–direct products of groups.

## 3. Presentation of a semi–direct product.

Recall, that if we have a group $G$ acting on the right on some other group $H$ by automorphisms, we can construct the (outer) semi–direct product $G \ltimes H$ as the set $G \times H$ with the composition rule

$$(g, h)(\tilde{g}, \tilde{h}) = (g\tilde{g}, (h\cdot\tilde{g})\tilde{h}) \ .$$

It is easy to see, that this composition makes $G \ltimes H$ into a group, and that $\overline{H} = e_G \times H$ is a normal subgroup of $G \ltimes H$ , isomorphic to $H$ .

Suppose now, that we have presentations of $G$ and $H$ given by the exact sequences

$$1 \to \langle \mathscr{R}^{F(X)} \rangle \xrightarrow{\subseteq} F(X) \xrightarrow{\lambda} G \to 1 \ ,$$
$$1 \to \langle \mathscr{S}^{F(Y)} \rangle \xrightarrow{\subseteq} F(Y) \xrightarrow{\mu} H \to 1 \ .$$

We want to find a presentation of $G \ltimes H$ , and to that end we define a homomorphism $\theta\colon F(X \cup Y) \to G \ltimes H$ by

$$\theta x = (\lambda x\,,\, e_H) \ , \ x \in X \ ,$$

$$\theta y = (e_G\,,\, \mu y) \ , \ y \in Y \ ,$$

(of course we assume, that X and Y are disjoint sets). We consider $F(X)$ and $F(Y)$ as subgroups of $F(X \cup Y)$ in the obvious way, and if $U \in F(X)$, $V \in F(Y)$, it is not hard to see, that

$$\theta(UV) = (\lambda U, \mu V).$$

This shows, that $\theta$ is surjective, so that the sequence

$$1 \to \text{Ker}\,\theta \xrightarrow{\subset} F(X \cup Y) \xrightarrow{\theta} G \ltimes H \to 1$$

is exact, and all that remains is to find a set of relators generating $\text{Ker}\,\theta$. The idea is the following: Given any $W \in F(X \cup Y)$, we should be able to find $U \in F(X)$ and $V \in F(Y)$, such that $\theta(W) = \theta(UV)$, for then the problem of deciding whether $W \in \text{Ker}\,\theta$ is reduced to the problem of deciding whether $U \in \text{Ker}\,\lambda$ and $V \in \text{Ker}\,\mu$, which can be done by means of the relators in $\mathcal{R}$ and $\mathcal{Q}$. If we look at the simple case $W = yx$, we have

$$\theta(yx) = (e_G, \mu y)(\lambda x, e_H) = (\lambda x, \mu y \cdot \lambda x) = (\lambda x, \mu M) = \theta(xM)$$

for some $M \in F(Y)$. Of course there may be several M's with $\mu M = \mu y \cdot \lambda x$, so for each pair $(x, y) \in X \times Y$ we will choose a particular one and denote it by $M(x, y)$. It is reasonable to expect, that knowing all the $M(x, y)$'s will enable us to "move the x's to the left" in any given $W \in F(X \cup Y)$, thereby solving our problem.

This is in fact true, so that letting

$$\mathcal{M} = \{M(x, y)^{-1} x^{-1} yx \mid x \in X, y \in Y\},$$

we have the

<u>Theorem.</u> $G \ltimes H$ *can be presented as* $\langle X \cup Y \mid \mathcal{R} \cup \mathcal{Q} \cup \mathcal{M} \rangle$.

The details of the proof are left to the reader, or can be found in the book by Johnson, noting that $G \ltimes H$ is an extension of G by H.

We remark that the subgroup $\langle \theta\{Y\} \rangle$ of $G \ltimes H$ is $\overline{H}$, which, as noted earlier, is normal and isomorphic to H.

Now, the abstract braid group $B_n$ can be made to act on the right by automorphisms on the free group $F(y_1,...,y_n)$ , namely by defining

$$y_k \cdot \lambda\sigma_j = \begin{cases} y_k & \text{if } k \neq j,\ j+1 \\ y_j\, y_{j+1}\, y_j^{-1} & \text{if } k = j \\ y_j & \text{if } k = j+1 \end{cases}$$

(cf. p. 25). Hence $B_n \ltimes F(y_1,...,y_n)$ has a presentation with

generators: $\sigma_1,...,\sigma_{n-1}$ , $y_1,...,y_n$

relations:
$$\sigma_j\, \sigma_k = \sigma_k\, \sigma_j \ , \ k > j+1 ,$$
$$\sigma_j\, \sigma_{j+1}\, \sigma_j = \sigma_{j+1}\, \sigma_j\, \sigma_{j+1} \ ,$$
$$\sigma_j^{-1}\, y_k\, \sigma_j = y_k \ , \ k \neq j , j+1 ,$$
$$\sigma_j^{-1}\, y_j\, \sigma_j = y_j\, y_{j+1}\, y_j^{-1} \ ,$$
$$\sigma_j^{-1}\, y_{j+1}\, \sigma_j = y_j \ .$$

Comparing this with the presentation of $D_n$ found in the previous section, we immediately have the

Theorem (Chow). $D_n \cong B_{n-1} \ltimes F(a_1,...,a_{n-1})$ .

Furthermore, the subgroup $U_n = \langle \theta a_1,...,\theta a_{n-1} \rangle$ of $D_n$ is normal and isomorphic to $F(a_1,...,a_{n-1})$ , and therefore free (cf. p. 24).

## 4. A presentation of $H_n$ .

The isomorphism $\zeta\colon B_{n-1} \ltimes F(a_1,...,a_{n-1}) \to D_n$ established in the previous section is explicity given by

$$\zeta(\beta, 1) = \beta \ , \ \zeta(\text{trivial braid},\ a_k) = \lambda(\sigma_{n-1} \cdots \sigma_{k+1}\, \sigma_k^2\, \sigma_{k+1}^{-1} \cdots \sigma_{n-1}^{-1}) \ ,$$

where we consider $B_{n-1}$ as a subgroup of $B_n$ in the obvious way. From this we see that the diagram

$$\begin{array}{ccc} B_{n-1} \ltimes F(a_1,\dots,a_{n-1}) & \xrightarrow[\cong]{\zeta} & D_n \\ & \pi \searrow \quad \swarrow \tau & \\ & \Sigma_n & \end{array}$$

commutes, where $\pi(\beta, V) = \tau\beta$. Hence by restriction $\zeta$ gives an isomorphism

$$H_{n-1} \ltimes F(a_1,\dots,a_{n-1}) = \operatorname{Ker} \pi \cong \operatorname{Ker} \tau = H_n \ .$$

Using recursion, this will enable us to find a presentation of $H_n$. Let $H_s$, $2 \leq s \leq n$, have a presentation given by the exact sequence

$$1 \to \langle \mathcal{L}_s^{F(C_s)} \rangle \xrightarrow{\subseteq} F(C_s) \xrightarrow{\gamma_s} H_s \to 1 \ ,$$

and let $\zeta_s\colon H_{s-1} \ltimes F(A_s) \to H_s$, $A_s = \{a_{1s},\dots,a_{(s-1)s}\}$, be the isomorphism mentioned above. From the previous section we know that $H_{s-1} \ltimes F(A_s)$ has a presentation given by

$$1 \to \langle (\mathcal{L}_{s-1} \cup \mathcal{M}_s)^{F(C_{s-1} \cup A_s)} \rangle \xrightarrow{\subseteq} F(C_{s-1} \cup A_s) \xrightarrow{\nu_s} H_{s-1} \ltimes F(A_s) \to 1 \ ,$$

where

$$\mathcal{M}_s = \{(a_{rs} \cdot \gamma_{s-1}\, c)^{-1}\, c^{-1}\, a_{rs}\, c \mid c \in C_{s-1} \ , \ 1 \leq r < s\} \ ,$$

and $\nu_s$ is given by

$$\nu_s c = (\gamma_{s-1} c, 1) \ , \ \nu_s a_{rs} = (\text{trivial braid}, a_{rs}) \ .$$

A natural choice is therefore to put

$$C_s = C_{s-1} \cup A_s \ , \ \mathcal{L}_s = \mathcal{L}_{s-1} \cup \mathcal{M}_s \ ,$$

and let $\gamma_s = \zeta_s \circ \nu_s$, i.e.

$$\gamma_s c = \zeta_s(\gamma_{s-1} c, 1) = \gamma_{s-1} c \ ,$$

$$\gamma_s a_{rs} = \zeta_s(\text{trivial braid}, a_{rs}) = \lambda(\sigma_{s-1} \cdots \sigma_{r+1}\, \sigma_r^2\, \sigma_{r+1}^{-1} \cdots \sigma_{s-1}^{-1}) \ .$$

Since $H_2 = D_2 \cong F(A_2)$ , we have

$$C_2 = A_2 \ , \ \mathscr{L}_2 = \emptyset \ , \ \gamma_2(a_{12}) = \lambda\sigma_1^2 \ ,$$

and recursion now shows that

$$C_n = A_2 \cup \ldots \cup A_n = \{a_{rs} \mid 1 \leq r < s \leq n\} \ ,$$
$$\mathscr{L}_n = \mathscr{M}_3 \cup \ldots \cup \mathscr{M}_n = \{(a_{ij} \cdot \gamma_n a_{rs})^{-1} a_{rs}^{-1} a_{ij} a_{rs} \mid 1 \leq i < j \leq n , 1 \leq r < s < j\},$$
$$\gamma_n a_{rs} = \lambda(\sigma_{s-1} \cdots \sigma_{r+1} \sigma_r^2 \sigma_{r+1}^{-1} \cdots \sigma_{s-1}^{-1}) \ .$$

Of course this can be formally proven by induction. Thus, in order to complete our task we only have to work out

$$a_{ij} \cdot \lambda(\sigma_{s-1} \cdots \sigma_{r+1} \sigma_r^2 \sigma_{r+1}^{-1} \cdots \sigma_{s-1}^{-1})$$

for all $1 \leq i < j \leq n$ , $1 \leq r < s < j$ , using

$$a_{ij} \cdot \lambda\sigma_k = a_{ij} \quad \text{if } i \neq k , k+1 \ ,$$
$$a_{kj} \cdot \lambda\sigma_k = a_{kj} \, a_{(k+1)j} \, a_{kj}^{-1} \ ,$$
$$a_{(k+1)j} \cdot \lambda\sigma_k = a_{kj} \ ,$$

and

$$a_{ij} \cdot \lambda\sigma_k^{-1} = a_{ij} \quad \text{if } i \neq k , k+1 \ ,$$
$$a_{kj} \cdot \lambda\sigma_k^{-1} = a_{(k+1)j} \ ,$$
$$a_{(k+1)j} \cdot \lambda\sigma_k^{-1} = a_{(k+1)j}^{-1} \, a_{kj} \, a_{(k+1)j} \ ,$$

(cf. p. 25 and 29). The results are as follows:

| | $a_{ij} \cdot \lambda(\sigma_{s-1} \cdots \sigma_{r+1}\, \sigma_r^2\, \sigma_{r+1}^{-1} \cdots \sigma_{s-1}^{-1})$ |
|---|---|
| $r < s < i < j$ | $a_{ij}$ |
| $r < s = i < j$ | $a_{rj}\, a_{ij}\, a_{rj}^{-1}$ |
| $r < i < s < j$ | $a_{rj}\, a_{sj}\, a_{rj}^{-1}\, a_{sj}^{-1}\, a_{ij}\, a_{sj}\, a_{rj}\, a_{sj}^{-1}\, a_{rj}^{-1}$ |
| $r = i < s < j$ | $a_{rj}\, a_{sj}\, a_{ij}\, a_{sj}^{-1}\, a_{rj}^{-1}$ |
| $i < r < s < j$ | $a_{ij}$ |

We illustrate the computations in the case $r < s = i < j$ :

$$\begin{aligned}
a_{ij} \cdot \lambda(\sigma_{i-1} \cdots \sigma_{r+1}\, \sigma_r^2\, \sigma_{r+1}^{-1} \cdots \sigma_{i-1}^{-1}) &= a_{(i-1)j} \cdot \lambda(\sigma_{i-2} \cdots \sigma_{r+1}\, \sigma_r^2\, \sigma_{r+1}^{-1} \cdots \sigma_{i-1}^{-1}) \\
&= \ldots \\
&= a_{(r+1)j} \cdot \lambda(\sigma_r^2\, \sigma_{r+1}^{-1} \cdots \sigma_{i-1}^{-1}) \\
&= a_{rj} \cdot \lambda(\sigma_r\, \sigma_{r+1}^{-1} \cdots \sigma_{i-1}^{-1}) \\
&= (a_{rj}\, a_{(r+1)j}\, a_{rj}^{-1}) \cdot \lambda(\sigma_{r+1}^{-1} \cdots \sigma_{i-1}^{-1}) \\
&= (a_{rj}\, a_{(r+2)j}\, a_{rj}^{-1}) \cdot \lambda(\sigma_{r+2}^{-1} \cdots \sigma_{i-1}^{-1}) \\
&= \ldots \\
&= a_{rj}\, a_{ij}\, a_{rj}^{-1} .
\end{aligned}$$

Collecting the above, we finally have the

<u>Theorem.</u> *The abstract coloured braid group* $H_n$ *has a presentation with*

*generators:* $a_{rs}$ , $1 \leq r < s \leq n$ ,

*relations:*

$$a_{rs}^{-1}\, a_{ij}\, a_{rs} = \begin{cases} a_{ij} & r < s < i < j \text{ or } i < r < s < j \\ a_{rj}\, a_{ij}\, a_{rj}^{-1} & r < s = i < j \\ a_{rj}\, a_{sj}\, a_{ij}\, a_{sj}^{-1}\, a_{rj}^{-1} & r = i < s < j \\ a_{rj}\, a_{sj}\, a_{rj}^{-1}\, a_{sj}^{-1}\, a_{ij}\, a_{sj}\, a_{rj}\, a_{sj}^{-1}\, a_{rj}^{-1} & r < i < s < j \end{cases}$$

Math. Proc. Camb. Phil. Soc. 99(1986), 247–260.
Reprinted with the permission of H.R. Morton.

APPENDIX 2

# Threading knot diagrams

BY H. R. MORTON

*Department of Pure Mathematics, University of Liverpool L69 3BX*

(*Received* 15 *February* 1985; *revised* 11 *July* 1985)

## 1. *Introduction*

Alexander [1] showed that an oriented link $K$ in $S^3$ can always be represented as a closed braid. Later Markov [5] described (without full details) how any two such representations of $K$ are related. In her book [3], Birman gives an extensive description, with a detailed combinatorial proof of both these results.

In this paper I shall describe a simple method of representing an oriented link $K$ as a closed braid, starting from a knot diagram for $K$ and 'threading' a suitable unknotted curve $L$ through the strings of $K$ so that $K$ is *braided* relative to $L$, i.e. $K \cup L$ forms a closed braid together with its axis.

I shall then give a straightforward derivation of Markov's result, using the ideas of threading, and a geometric version of the braid moves with which Markov relates two braids representing the same $K$. The geometric approach is described in terms of links $K \cup L$, in which $K$ forms a closed braid relative to an axis $L$. Such a link will be called *braided*, and in addition it will be called a *threading* of an explicit diagram for $K$ if it arises from the threading construction. Two braided links which are related by the geometric version of Markov's moves will be called *Markov-equivalent*. Markov's theorem, which says in this geometric translation that braided links $K \cup L$ and $K' \cup L'$ are Markov-equivalent if and only if the oriented links $K$ and $K'$ are isotopic, will then follow from Theorems 2, 3 and 4, on threadings.

These results on threadings, whose proofs are not elaborate, are as follows:

THEOREM 2. *Any braided link $K \cup L$ arises as a threading of some diagram for $K$.*

THEOREM 3. *Any two threadings of a given diagram of $K$ are Markov-equivalent.*

THEOREM 4. *Two different diagrams of $K$ have Markov-equivalent threadings.*

*Remarks*. Bennequin [2] gives a geometric proof of Markov's theorem in the course of his work on contact structures, using suitably positioned spanning surfaces for the closed braids; see also Rudolph [7] for a discussion of such surfaces.

Markov's result itself, and also the representation of a knot as a closed braid, have had attention recently following Jones' use of braid groups in constructing his new polynomial knot-invariant [4].

## 2. *Notation and definitions*

For definitions and notation concerning braids I shall refer to [3]. In particular, given an $n$-string braid $\beta \in B_n$ I shall refer to the *closure* of $\beta$, written $\hat{\beta}$, to mean an oriented link which arises from an explicit geometric representative of $\beta$ in $D^2 \times I$ by identifying the ends of the cylinder. The closure of $\beta$ is determined by $\beta$ up to isotopy in $S^3$. In fact $\beta$ determines up to isotopy a link $\hat{\beta} \cup L_\beta$ consisting of its closure $\hat{\beta}$, lying in the complement of $L_\beta$, an unknotted curve called the *axis* of $\hat{\beta}$; here the exterior of $L_\beta$ is an unknotted solid torus $D^2 \times S^1$ in which the curve $\hat{\beta}$ lies regularly with respect to the projection to $S^1$. I shall refer to the link $\hat{\beta} \cup L_\beta$, with a natural choice of orientation, as the *complete closure* of $\beta$.

If an oriented link $K \cup L$ is given in which $L$ is unknotted, and $K$ projects regularly to $S^1$ under some choice of product projection $p_L: S^3 - L \to S^1$ then I call $K \cup L$ *braided* (relative to $L$). Then the complete closure of $\beta$ is braided, and conversely any braided link is the complete closure of some $\beta$. Geometrically the complete closure captures $\beta \in B_n$ very well, for $\beta$ and $\gamma$ have isotopic complete closures (respecting orientation) if and only if $\beta$ and $\gamma$ are conjugate in $B_n$, see e.g. [6].

Consequently a braided link $K \cup L$ determines $\beta \in B_n$, ($n = \mathrm{lk}(K, L)$), up to conjugacy.

A TEST FOR A BRAIDED LINK. *If $L$ is unknotted, and a product projection*

$$p_L: S^3 - L \to S^1$$

*is found in which $K$ is mapped monotonically, i.e. for a suitable orientation of $S^1$ the map $p_L$ is locally increasing on $K$, but not necessarily* strictly *increasing, and in addition $p_L$ is not constant on any component of $K$, then $K \cup L$ is braided.*

*Proof.* Under these conditions an arbitrarily small isotopy of $K$ can be made in $S^3 - L$ to ensure that $p_L$ becomes strictly increasing.

*Markov moves*

A Markov move replaces a braid $(\beta, n) \in B_n$ by

(1) $(\gamma, n)$, with $\gamma$ conjugate to $\beta$ in $B_n$, or

(2) $(\beta\sigma_n^{\pm 1}, n+1) \in B_{n+1}$, or

(3) $(\beta', n-1) \in B_{n-1}$, if $\beta = \beta'\sigma_{n-1}^{\pm 1}$, and $\beta'$ is a word in $\sigma_1, \ldots, \sigma_{n-2}$.

The complete closures of two braids related by a type (1) move are isotopic; conversely we have noted that a braided link determines a braid up to type (1) moves.

I shall shortly describe a relation, *simple Markov-equivalence*, between two braided links which will ensure that they are complete closures of two braids related by a move of type (2), or its inverse, of type (3).

If *Markov-equivalence* is defined as the relation on braided links generated by isotopy and simple Markov-equivalence we have then the geometric reformulation of Markov's theorem which follows.

THEOREM 5. *If $\beta$ and $\gamma$ are two braids whose closures are isotopic as oriented links, then their complete closures are Markov-equivalent.*

*Definition.* Two braided links $K \cup L$ and $K' \cup L'$ are *simply Markov-equivalent* if the second one can be isotoped so that $L' = L$ and $K'$ agrees with $K$ except for arcs $\alpha$ of $K$ and $\alpha'$ of $K'$ with the following properties:

(1) we can find a projection $p_L: S^3 - L \to S^1$ which is constant on $\alpha$, strictly monotone on the rest of $K$ and monotone of degree 1 on $\alpha'$.

(2) there is a disc $A$ spanning $\alpha \cup \alpha'$ whose interior meets $L$ transversely in one point and avoids $K \cup \alpha'$.

From the definition it is clear that if $\mathrm{lk}(K, L) = n$ then $\mathrm{lk}(K', L') = n+1$, so the braided links will be complete closures of braids in $B_n$, $B_{n+1}$ respectively.

LEMMA 1. *If $K \cup L$ and $K' \cup L'$ are simply Markov-equivalent braided links then they are the complete closures of some $\beta \in B_n$ and $\beta\sigma_n^{\pm 1} \in B_{n+1}$ respectively where $n = \mathrm{lk}(K, L)$.*

*Proof*. Suppose that the links have been isotoped to agree except on arcs $\alpha$ of $K$ and $\alpha'$ of $K'$ as specified in the definition, and suppose that $\alpha$ lies at the level $p_L = \theta_0$. Look at the way in which the disc $A$ bounded by $\alpha \cup \alpha'$ meets the level disc $D$, with boundary $L$, for the level $p_L = \theta_0$.

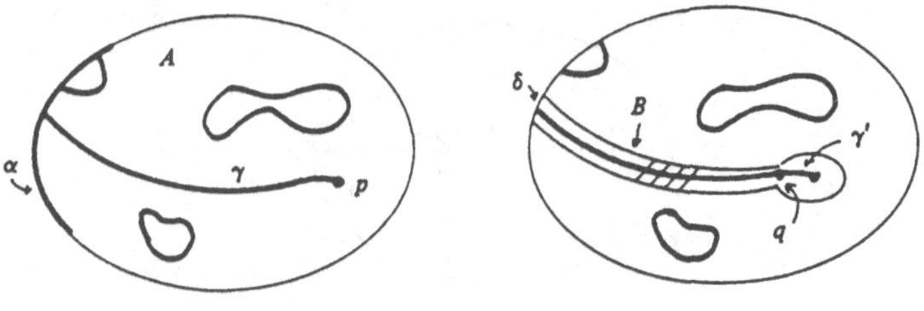


Fig. 1

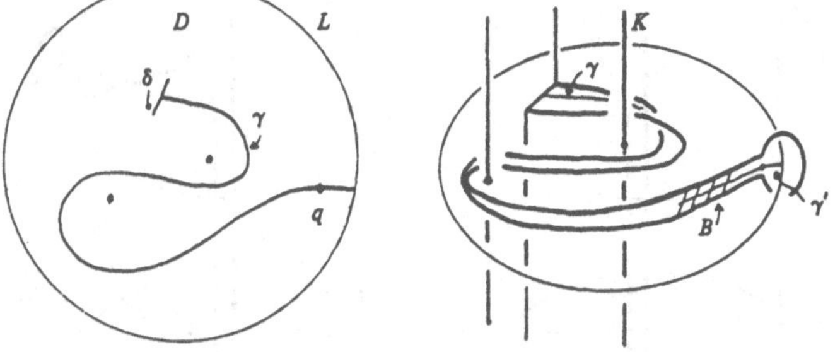


Fig. 2

After a slight isotopy of the interior of $A$ we may assume that it meets $D$ transversely and that $A \cap D$ consists of the arc $\alpha$ in $\partial A$ together with a finite number of disjoint closed curves and arcs as illustrated in Fig. 1. Where an arc meets $\alpha$ we may assume that $p_L$ behaves locally on $A$ like the restriction to one side of a saddle point.

The single point, $p$, of transverse intersection of $A$ with $L = \partial D$ will be the end-point of exactly one arc $\gamma$ of $A \cap D$, whose other end must lie on $\alpha$. Choose a small disc in $A$, centre $p$, which we can assume (after isotopy of $A$) to lie, except for $p$, as a product of a subarc $\gamma' \subset \gamma$ with $S^1$ in the solid torus $S^3 - L$. The boundary of this disc will eventually form a single braid string over the end-point $q$ of $\gamma'$ in $D$.

Adjoin to this disc a thin ribbon of $A$ about $\gamma$ small enough to contain no critical points for $p_L$ and to have $p_L$ monotone on its edge. This ribbon, whose end on the arc $\alpha$ will be called $\delta$, may be chosen to lie arbitrarily close to $\gamma$, say within the levels $[\theta_0 - \epsilon, \theta_0 + \epsilon]$ of $p_L$. Together with the disc about $p$ it makes up a disc $B \subset A$. We now use the isotopy determined by $A$ from $\partial A$ to $\partial B$ in the complement of $L$ to isotop $K$ and $K'$ into new positions where they are related by the disc $B$ in place of $A$. The disc $D$ and its relation with $K$, $K'$ and $B$ are illustrated in Fig. 2.

To make an explicit correspondence of $K$, $K'$ with the closure of two braids, choose $n+1$ reference points $q_1, \ldots, q_n, q$ in $D$, and a standard arc from $q_n$ through $q$ to $p$ on $\partial D$, which extends the arc $\gamma'$ from $q$ to $p$. These points (with or without $q$) will be the starting and finishing points for the braids on $n$ (or $n+1$) strings in $D \times I$ which become $K$, or $K'$, when closed.

We now complete the proof by isotoping $D$, keeping $L = \partial D$ and $\gamma'$ fixed so that $\alpha$ becomes the standard arc from $q_n$ to $p$, $\delta$ becomes a small arc through $q_n$ and $K$ meets $D$ in this small arc, together with the points $q_1, \ldots, q_{n-1}$. Extend this by a level preserving

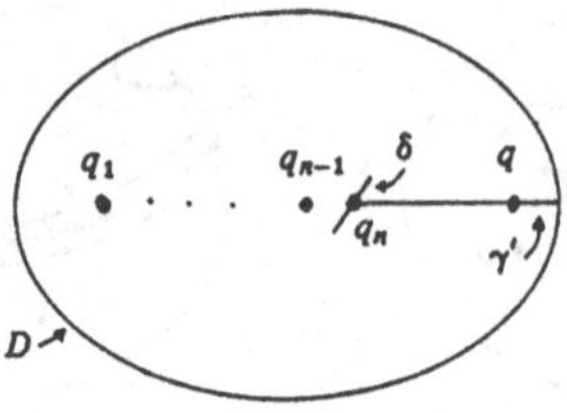


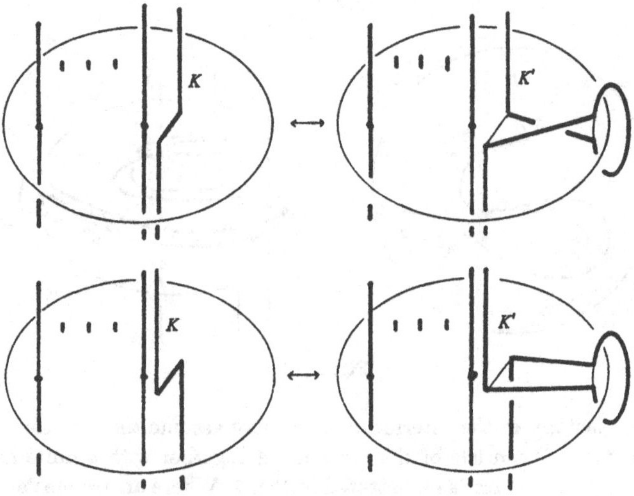


Fig. 3

isotopy which is the identity outside the levels $[\theta_0 \ldots 2\epsilon, \theta_0 + 2\epsilon]$ of $p_L$ and carries $B$ to a ribbon within these levels lying close to the standard arc together with the unchanged disc transverse to $L$ determined by $\gamma'$. Assume that the ribbon lies within the levels $[\theta_0 - \epsilon, \theta_0 + \epsilon]$, and that $K$ passes through $q_1, \ldots, q_{n-1}$ vertically between these levels, so that $K'$ differs from $K$ after this isotopy only by the addition of the ribbon edges close to level $\theta_0$ between $q_n$ and $q$ and an extra straight string above $q$ in the levels beyond the ribbon.

A slight adjustment to $K$ and $K'$ is still strictly necessary to realise both as the closure of braids based on the reference points when cut open at level $\theta_0$, or better at $\theta_0 - \epsilon$.

Fig. 3 shows the difference of $K$ and $K'$ between levels $\theta_0 - \epsilon$ and $\theta_0 + \epsilon$ for the two possible orientations of $\delta$ in $D$. It is then clear that the braids given by cutting open at $\theta_0 - \epsilon$ differ simply by the addition of an extra string about $q$ and a generator $\sigma_n^{\pm 1}$ which comes from the string exchange close to the standard arc joining $q_n$ to $q$. ∎

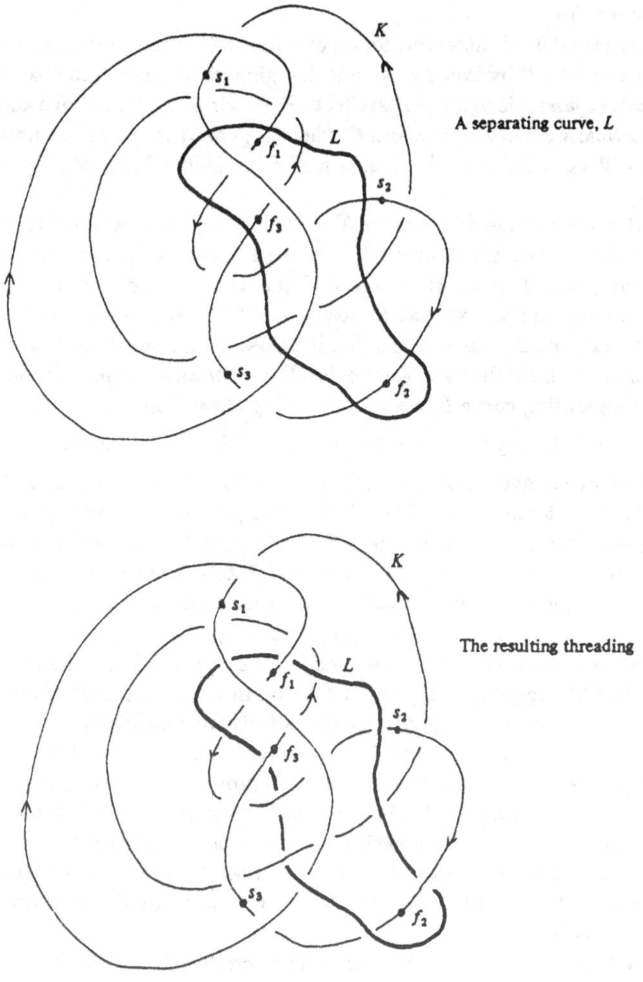


Fig. 4

## 3. *Threadings*

Starting from a diagram for an oriented link $K$ I shall describe how to find unknotted curves $L$ resulting in a variety of braided links $K \cup L$ which will be called *threadings* of the diagram for $K$.

By a diagram for $K$ I shall mean a simple projection of $K$ to some plane $P$ in which a finite number of over-crossing and under-crossing points in $K$ are distinguished – these are the inverse images of the double-points of the projection.

*Definition.* A *choice of overpasses* for a diagram will consist of $(S, F)$, two finite subsets $S = \{s_1, \ldots, s_k\}$, $F = \{f_1, \ldots, f_k\}$ of $K$ forming the 'starting' and 'finishing' points of overpasses, which alternate in $K$, and which divide $K$ into arcs of the form $[s, f]$ called *overpasses* containing no undercrossings, and $[f, s]$, *underpasses*, containing no overcrossings.

There is no need, however, for an overpass to contain any overcrossing.

We can alter $K$ by isotopy without changing its projection to $P$ so that its overpasses and underpasses lie in planes parallel to $P$, (horizontal) just above and below it, except for vertical segments near $S$ and $F$, where $K$ goes upwards or downwards respectively.

It will be helpful to think of $S$ and $F$ as points lying in $P$, in the projection of $K$.

A *threading* of the diagram for $K$ with the given choice of overpasses $(S, F)$ is constructed from any closed curve $L$ in $P$ which separates $S$ and $F$ as follows:

Arrange that $L$ crosses the image of $K$ transversely, and alter $K$ in the neighbourhood of each crossing point so that $K$ crosses *over* $L$ if it is passing from the side of $L$ which contains $S$, and $K$ crosses *under* $L$ if it passes from the side of $L$ which contains $F$, to give a link $K \cup L$ called a *threading*. See Fig. 4 for an example of a choice of overpasses and a separating curve $L$, with the resulting threading.

THEOREM 1. *Any threading of a diagram for K is a braided link.*

*Proof*. Select overpasses $(S, F)$ for the diagram, and a curve $L$ in the plane $P$ of the diagram which separates $S$ from $F$. Now straighten out $L$ in the plane $P$ by an isotopy of $P$, carrying the projected image of $K$ along, so that $L$ becomes (almost) a straight line, with points of $S$ lying to one side and $F$ to the other. We can suppose that $K$ is isotoped at the same time so that the overpasses and underpasses lie in planes parallel to $P$, just above and below their projected image.

We now change our point of view, and imagine that $P$ forms the $xz$-plane and $L$ forms the $z$-axis (having sent one point of $L$ to infinity on $P$). Using polar coordinates based on $L$ as axis, the plane $P$ splits into two half-planes, one, given by $\theta = 0$ say, containing the points of $F$, and the other, $\theta = \pi$, containing the points of $S$.

Project the overpasses of $K$ to the half-planes $\theta = -\epsilon$ and $\theta = \pi + \epsilon$, and similarly project the underpasses to the half-planes $\theta = \epsilon$ and $\theta = \pi - \epsilon$, for some suitably small $\epsilon$. These curves (which cross the axis $L$ in various points corresponding to the crossings of $L$ with the projected image of $K$) are then joined up by vertical arcs (i.e. in the direction of projection) through the points of $S$ and $F$ to give a closed curve isotopic to $K$ with the same projected image.

Apart from the points where this curve crosses $L$ the polar coordinate increases monotonically, since it is constant, except on the vertical arcs, where it increases from $-\epsilon$ to $\epsilon$ for those through a point of $F$, because of moving from an overpass to an underpass, and it increases from $\pi - \epsilon$ to $\pi + \epsilon$ for the arcs through points of $S$.

As illustrated in Figs. 5 and 6, which show the process for a simple knot diagram, the threading construction now diverts the curve $K$ near its crossings with $L$ to run around $L$ in the direction of increasing $\theta$. We can arrange explicitly that pieces of $K$ which pass through the cylinder $r \leqslant \delta$ are diverted to run around $r = \delta$, with pieces of $K$ which come from the $S$-side $(\theta = \pi \pm \epsilon)$ being taken above $L$ (i.e. through $\theta = \frac{1}{2}\pi$) and pieces which come from the $F$-side $(\theta = \pm\epsilon)$ being taken beneath. Thus, where, say, an overpass crosses $L$ from side $S$ to side $F$ the polar coordinate after threading will increase by less than $\pi$, from $\pi + \epsilon$ to $-\epsilon$, while on an overpass which crosses from side $F$ to side $S$ the increase will be $\pi + 2\epsilon$.

Consequently the threading has been isotoped so that, with $L$ as axis, the curve $K$

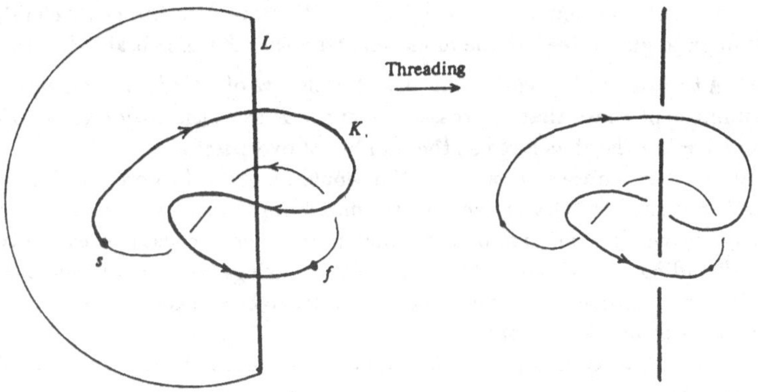
L
Threading
K.
s
f

Fig. 5

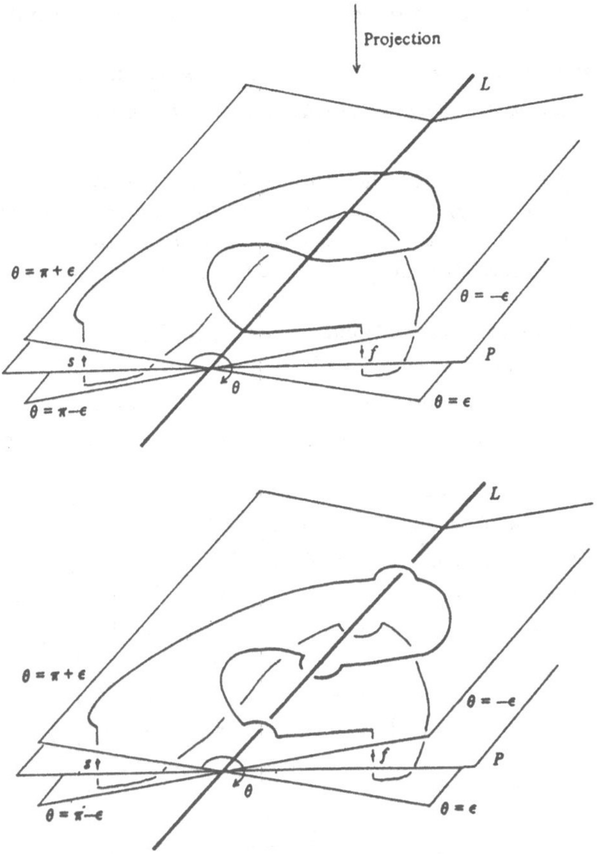
Projection
L
θ = π + ε
θ = −ε
P
s
f
θ
θ = π−ε
θ = ε
L
θ = π + ε
θ = −ε
P
s
f
θ
θ = π−ε
θ = ε

Fig. 6

runs monotonically (althought not strictly so) with respect to the polar coordinate. This is enough to guarantee, by the monotone test, that $K \cup L$ is braided. ∎

*Remark.* A threading of $K$ will exhibit $K$ as the closure of a braid on $n$ strings, where $n$ is the number of times that $K$ crosses the curve $L$ from the $S$-side to the $F$-side; consequently $n$ is at least as large as the number of overpasses.

In an attempt to reduce $n$ for a given $K$ it would be natural to present $K$ so that a small number of overpasses can be chosen (the minimum possible is the bridge number). It is clearly always possible to find an $L$ which meets each overpass exactly once; the resulting threading may however have many more strings, for there is no guarantee that the underpasses all cross $L$ once, and $L$ will have to be threaded under them every time they cross from side $S$ to side $F$.

Nevertheless the threading process is a very quick means of exhibiting any $K$ as a closed braid, starting from any diagram of $K$.

*Alexander's theorem*, that any oriented link $K$ can be represented as a closed braid, is an immediate corollary of Theorem 1, since any diagram of $K$ can be threaded in many ways.

I shall now show that all braided links arise from threadings in the next Theorem.

THEOREM 2. *The complete closure of a braid $\beta \in B_n$ is a threading of some diagram of its closure $\hat{\beta}$.*

*Proof.* A braid $\beta$ corresponds to a homeomorphism $h$ of the disc $D^2$ leaving $\partial D^2$ fixed and $n$ points $p_1, \ldots, p_n$ invariant. In this correspondence we choose an isotopy of $h$ to the identity, rel $\partial D^2$, to give a level-preserving homeomorphism $H: D^2 \times I \to D^2 \times I$, with $h = H|D^2 \times \{1\}$. The image of $\{p_1, \ldots, p_n\} \times I$ form the strings of a representative braid for $\beta$.

Choose disjoint arcs $a_1, \ldots, a_n$ in $D^2$ joining points $r_i \in \partial D^2$ to $p_i$. The closure of $\beta$ is then isotopic to $H(\{p_1, \ldots, p_n\} \times I)$ together with

$$(a_1 \cup \ldots \cup a_n) \times \{0\} \cup (a_1 \cup \ldots \cup a_n) \times \{1\} \cup \{r_1, \ldots, r_n\} \times I,$$

and the axis of $\hat{\beta}$ can be represented by a circle just inside $\partial D^2 \times \{\frac{1}{2}\}$; as shown in Fig. 7.

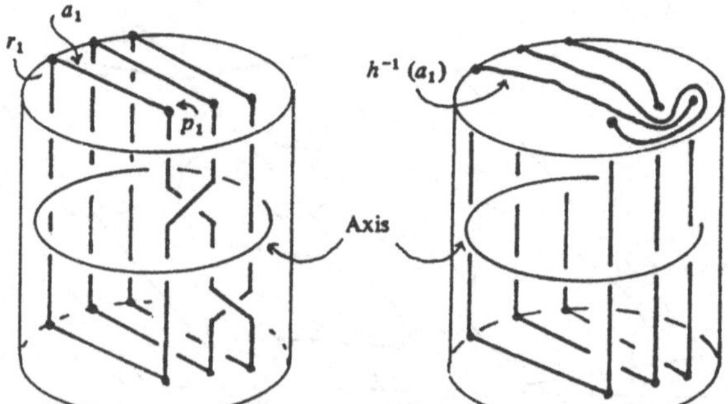


Fig. 7

Apply the homeomorphism $H^{-1}$ to $D^2 \times 1$ to show that $\hat{\beta}$ is isotopic to the vertical lines $\{p_1, \ldots, p_n\} \times I \cup \{r_1, \ldots, r_n\} \times I$ together with the arcs $a_1 \cup \ldots \cup a_n$ in level 0 overlaid with the arcs $h^{-1}(a_1) \cup \ldots \cup h^{-1}(a_n)$ in level 1. The axis remains as before, and $\hat{\beta}$ is represented as the threading of a diagram in $D^2$ with $S = \{p_1, \ldots, p_p\}$, $F = \{r_1, \ldots, r_n\}$, underpasses $a_1, \ldots, a_n$ and overpasses $h^{-1}(a_1), \ldots, h^{-1}(a_n)$, where the threading curve is a circle just inside $\partial D^2$. ∎

## 4. *Markov's Theorem*

As shown by Lemma 1, Markov's theorem, which says that any two braids with isotopic closures are related by a sequence of Markov moves, can be given in geometric form as follows.

THEOREM 5. (Markov). *If $K \cup L$ and $K' \cup L'$ are braided links, and $K$ is isotopic to $K'$, as oriented links, then $K \cup L$ and $K' \cup L'$ are Markov-equivalent.*

The theorem will follow by using Theorem 2 to show that $K \cup L$ and $K' \cup L'$ are both threadings of some diagrams of $K$. Then Theorems 3 and 4 will complete the proof.

THEOREM 3. *Any two threadings of a given diagram for $K$ are Markov-equivalent.*

THEOREM 4. *Any two diagrams for $K$ have Markov-equivalent threadings.*

*Proof of Theorem* 5. By Theorem 2, $K \cup L$ is a threading of some diagram of $K$. Again by Theorem 2, $K' \cup L'$ is a threading of some diagram of $K'$; since $K'$ is isotopic to $K$ this is a threading of a second diagram of $K$.

By Theorem 4 we can choose threadings of the first and second diagrams of $K$ which are Markov-equivalent. By Theorem 3 the chosen threading of the first diagram is Markov-equivalent to $K \cup L$, since $K \cup L$ is another threading of the same diagram of $K$; the chosen threading of the second diagram is similarly Markov-equivalent to $K' \cup L'$. ∎

To prove Theorem 3 we show it first with a given choice of overpasses $(S, F)$, in Lemma 2. Independence of the choice of overpasses follows, using Lemma 3 to construct a choice of overpasses $(S'', F'')$ with $S, S' \subset S''$; $F, F' \subset F''$, for any two given choices $(S, F)$ and $(S', F')$. Then any threading of $(S'', F'')$ will give threadings of $(S, F)$ and $(S', F')$ which are isotopic.

For Theorem 4 is is enough to show that two diagrams of $K$ which differ by a Reidemeister move have isotopic, hence Markov-equivalent, threadings for some choice of overpasses. This is done by choosing $(S, F)$ and the threadings to be identical outside the region of the move, and only to involve the region very simply, if at all.

I now complete the proof of Theorem 5 by proving Lemmas 2 and 3, and Theorem 4. In the accompanying diagrams the curve $L$ to be threaded is drawn more thickly than $K$.

LEMMA 2. *Given a diagram for an oriented link $K$ in a plane $P$, with choice of overpasses $(S, F)$, then the threadings defined by any two simple closed curves $L$, $L'$ which separate $S$ and $F$ are Markov-equivalent.*

*Proof.* We may suppose, without loss of generality, that $F$ lies in the bounded component of $P - L$ and of $P - L'$.

(*a*) Suppose that $L$ and $L'$ are isotopic in $P - (S \cup F)$. Then $L$, $L'$ and $K$ are related by a sequence of moves of type 1 or type 2 shown in Fig. 8, in which no points of $S$ or $F$ appear.

The two threadings in type 2 are clearly isotopic, except when the orientations require them to look as in Fig. 9. A picture like that in Fig. 9 can indeed occur, as seen

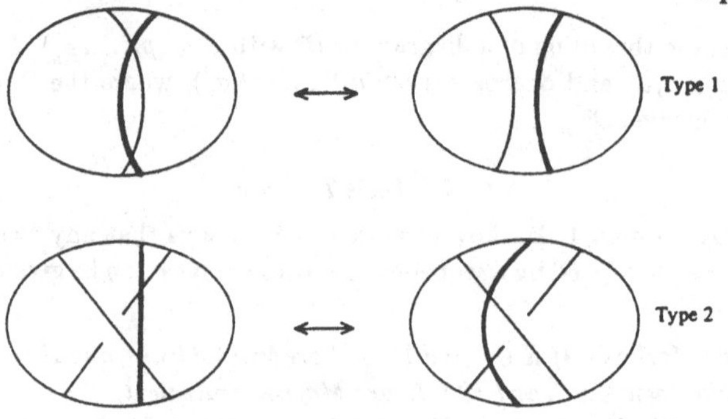


Fig. 8.

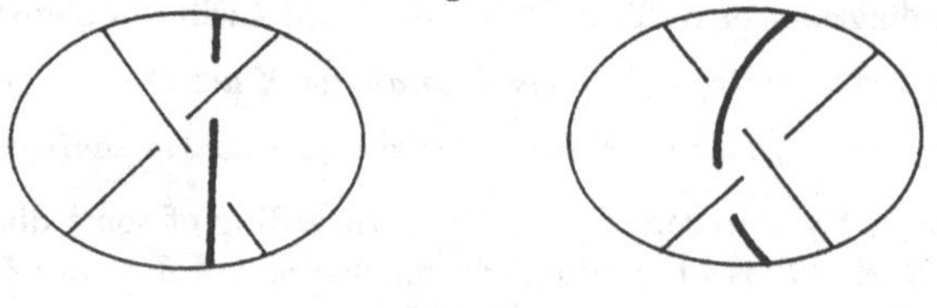

Fig. 9.

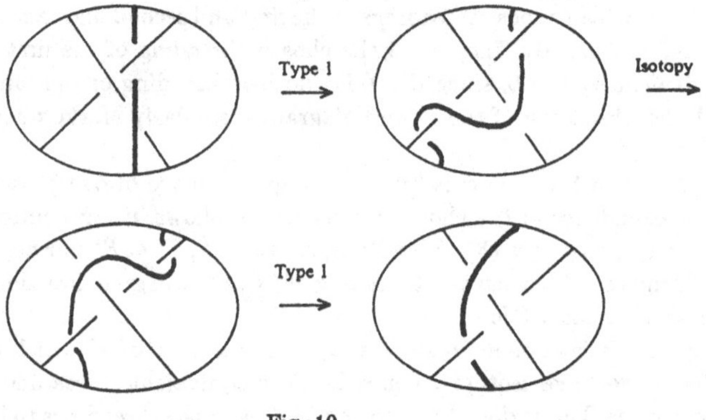


Fig. 10.

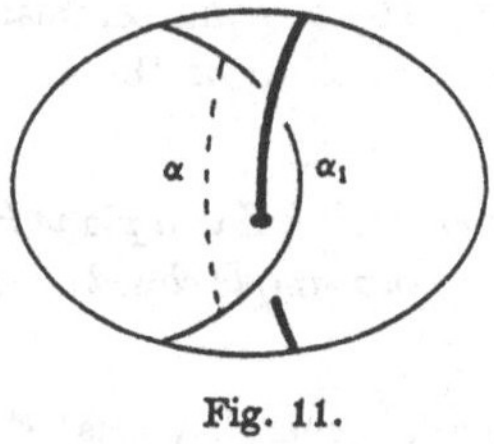


Fig. 11.

for example near the bottom of Fig. 4. In this case one threading converts to the other by a sequence of type 1 moves and isotopies, as in Fig. 10, noting that undercrossings and overcrossings of $K$ with $L$ must alternate on passing around $K$. It is then enough to show that the two threadings in a type 1 move are simply Markov-equivalent. Now the whole of $K$ in the diagram belongs either to one overpass or to one underpass. In the

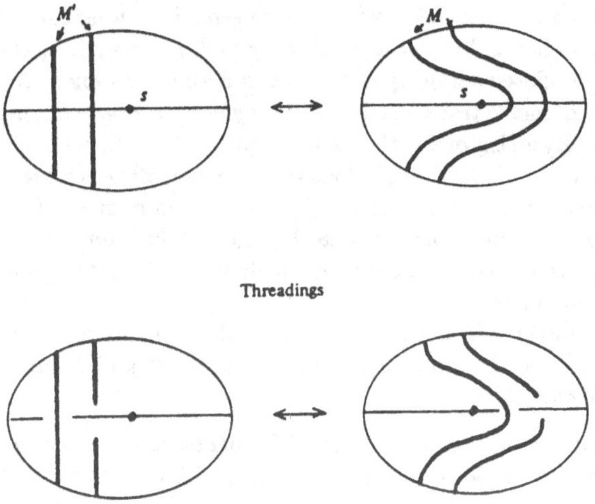


Fig. 12.

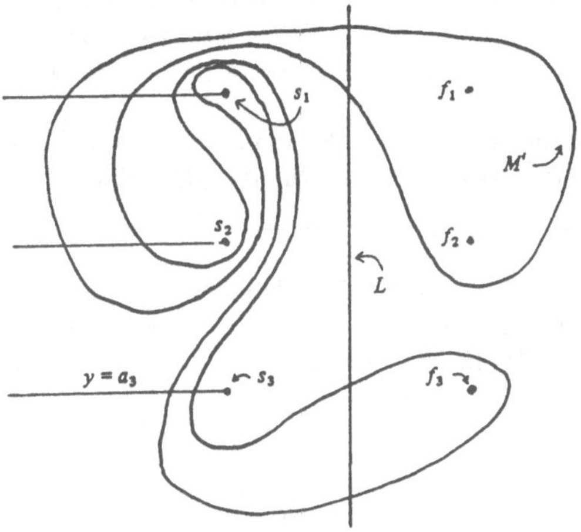


Fig. 13.

threading construction we may then assume that the two parts of $K$ to one side of $L$ lie at the same level of $p_L$ (it will be $\theta = \pi \pm \epsilon$ or $\pm\epsilon$ according to side and whether we have an under or overpass) until reaching the immediate neighbourhood of $L$. Join them by an arc $\alpha$ at this level, as indicated in Fig. 11. Then $\alpha$ together with the arc $\alpha'$ of $K$ crossing $L$ bounds a disc as required for a simple Markov-equivalence between the two threadings.

(*b*) To deal with the general case, observe that if a curve $M'$ in $P$ separating $S$ and $F$ is isotoped to $M$ by pushing a *pair* of crossings with $K$ past a point of $S$ (or of $F$) as in Fig. 12, then $M$, which also separates $S$ and $F$, and $M'$ define isotopic threadings.

By isotopy of $P$ we may make an explicit choice of $S$, $F$ and $L$. Let us take $S$ and $F$ to consist of the points $\{(-1, a_i)\}$ and $\{(1, a_i)\}$ respectively, for some $a_1, \ldots, a_k$, and take

$L$ to be part of the $y$-axis, completed with a large semicircle to enclose $F$. We may suppose also that $K$ runs parallel to the $x$-axis near each point $s_1, \ldots, s_k$ of $S$.

Let $L'$ be any simple closed curve separating $S$ and $F$ and enclosing $F$. Assume, after a small isotopy that $L'$ meets transversely the lines $y = a_i$, $x < -1$ running from $s_i$ to infinity away from $L$. By an isotopy in the complement of $S$ and $F$, pushing along these lines, we can alter $L'$ to $M'$ whose intersections with the lines all lie close to $S$. A picture of a typical $M'$ is given in Fig. 13. Each line is met an even number of times by $M'$, since $S$ lies outside $M'$; we may now isotop $M'$ by moving the intersections across the points of $S$ a pair at a time to reach a curve $M$ which does not meet the lines, and so is isotopic to $L$ in the complement of $S$ and $F$.

The threadings defined by $M$ and $L$, and by $M'$ and $L'$, are Markov-equivalent, by ($a$); we have just observed that the threadings defined by $M$ and $M'$ are isotopic, so Lemma 2 is established. ∎

Lemma 3. *Given a diagram for $K$ and a choice $(S, F)$ of overpasses, and any point $s$ of $K$ not in $F$, we can make a new choice of overpasses $(\bar{S}, \bar{F})$ with $s \in \bar{S}$, $S \subset \bar{S}$ and $F \subset \bar{F}$.*

*Proof.* If $s$ lies on an overpass of $(S, F)$ then choose $f$ immediately before $s$ on $K$, so that $[f, s] \subset K$ contains no overcrossing point, and take $\bar{F} = F \cup \{f\}$, $\bar{S} = S \cup \{s\}$. The original overpass containing $s$ becomes separated into two by the new underpass $[f, s]$

If $s$ lies on an underpass then choose $f$ immediately after $s$ with no undercrossings in $[s, f]$, and take $\bar{F}$, $\bar{S}$ as before. Then $[s, f]$ becomes a new overpass, separating one underpass into two. ∎

A similar argument allows extension of $F$ by any $f \notin S$. Two choices $(S, F)$ and $(S', F')$ of overpasses can then be extended readily (provided that

$$S \cap F' = S' \cap F = \varnothing)$$

to a choice of overpasses $(S'', F'')$ with $S \cup S' \subset S''$ and $F \cup F' \subset F''$.

Clearly any curve $L$ in $P$ which separates $S''$ and $F''$ will also separate $S$ and $F$, and so $L$ determines a threading for $(S'', F'')$ and for $(S, F)$. These threadings are actually

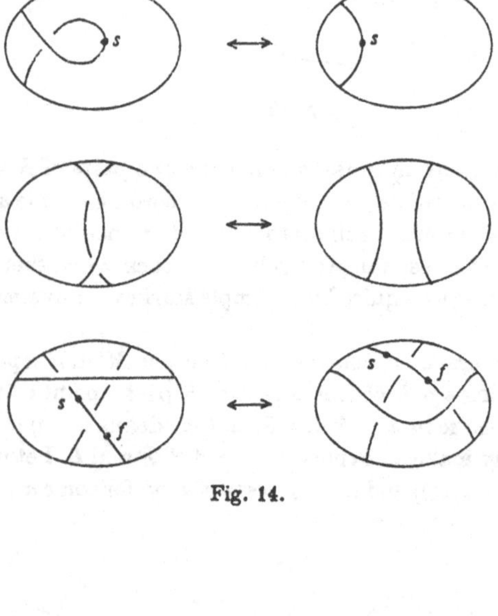


Fig. 14.

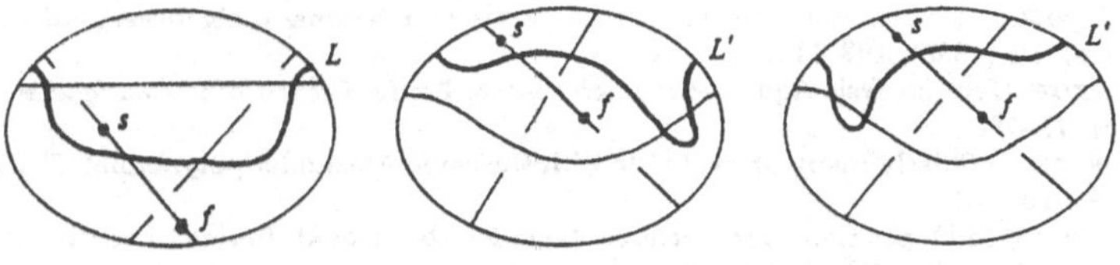


Fig. 15.

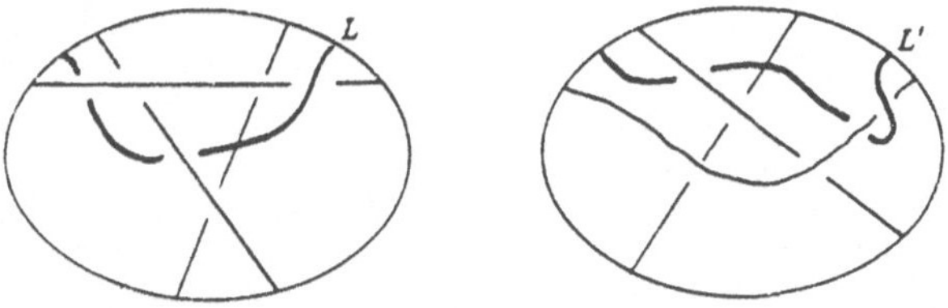


Fig. 16.

the same, since the construction of the threading from the diagram depends only on the sense in which $K$ crosses $L$. The dissection of $K$ into overpasses is required only to ensure that $L$ is a suitable curve to make $K \cup L$ braided. Consequently for any two choices of overpasses there is some common threading. The Markov-equivalence of any two threadings of a given diagram then follows using Lemma 2.

THEOREM 4. *Any two diagrams for an oriented link $K$ have Markov-equivalent threadings.*

*Proof.* Any two diagrams for $K$ are related by a sequence of Reidemeister moves, illustrated in Fig. 14, so it is enough to show how, for each Reidemeister move, isotopic threadings can be chosen for the two related diagrams.

By Theorem 3 we may make any convenient choice of overpasses. In choosing $(S, F)$ on any diagram we need only ensure that there is always a point of $S \cup F$ separating an overcrossing from a neighbouring undercrossing.

Place points $s \in S$ and $f \in F$ as indicated in Fig. 14, and choose the rest of $S \cup F$ to lie outside the region altered by the move, with $s$ and $f$ interchanged if the orientation of $K$ is in the opposite sense. In the case of the first two moves we can then choose our separating curve $L$ to lie outside the region of change, so that the resulting threadings are unaltered by the move.

For the third move we must choose $L$ to separate $s$ and $f$. Fig. 15 shows part of a suitable $L$ which gives a threading isotopic to that from $L'$ or $L''$, depending on the orientation of the uppermost piece of $K$. Fig. 16 shows the relevant parts of the threadings in one of the two cases, making it clear that they are isotopic. The crossing of $L$ with the lowest piece of $K$ will not affect the isotopy. The proof of Theorem 4, and so also of Markov's theorem, is complete. ∎

## REFERENCES

[1] J. W. ALEXANDER. A lemma on systems of knotted curves. *Proc. Nat. Acad. Sci. U.S.A.* 9 (1923), 93–95.

[2] D. BENNEQUIN. Entrelacements et équations de Pfaff. *Astérisque* 107–108 (1983), 87–161.

[3] J. S. BIRMAN. *Braids, Links and Mapping Class Groups. Ann. of Math. Stud.* 82 (1974).

[4] V. F. R. JONES. A polynomial invariant for knots via von Neumann algebras. *Bull. Amer. Math. Soc.* **12** (1985), 103–111.

[5] A. A. MARKOV. Über die freie Äquivalenz geschlossener Zöpfe. *Recueil Mathématique Moscou* **1** (1935), 73–78.

[6] H. R. MORTON. Infinitely many fibred knots with the same Alexander polynomial. *Topology* **17** (1978), 101–104.

[7] L. RUDOLPH. Special positions for surfaces bounded by closed braids. Preprint 1984, Box 251, Adamsville, Rhode Island.

# BIBLIOGRAPHY

In the following bibliography we only list items, which are explicitly mentioned in the text, or from which material is used. Many of the items mentioned contain extensive bibliographies.

## Books


[1] J.S. Birman: Braids, Links and Mapping Class Groups. Annals of Mathematics Studies 82, Princeton Univ. Press, 1975.

[2] G. Burde and H. Zieschang: Knots. Walter de Gruyter, Berlin – New York, 1985.

[3] F.R. DeMeyer and E.C. Ingraham: Separable algebras over commutative rings. Lecture Notes in Mathematics 81, Springer–Verlag, 1971.

[4] J. Dugundji: Topology. Allyn and Bacon, Boston MA, 1966.

[5] L. Gillman and M. Jerison: Rings of continuous functions. 2nd Ed.: Graduate Texts in Mathematics 43, Springer–Verlag, 1976.

[6] R.C. Kirby and L.C. Siebenmann: Foundational Essays on Topological Manifolds, Smoothings, and Triangulations. Annals of Mathematics Studies 88, Princeton Univ. Press, 1977.

[7] S. MacLane: Homology. Grundlehren der Mathematischen Wissenschaften 114, Springer–Verlag, 1963.

[8] W. Magnus, A. Karrass and D. Solitar: Combinatorial Group Theory: Presentations of groups in terms of generators and relations. Interscience Publishers, Wiley and Sons Inc., New York, 1966.

[9] W.S. Massey: Algebraic Topology: An Introduction. 4th Ed: Graduate Texts in Mathematics 56, Springer–Verlag, 1977.

[10] S. Moran: The Mathematical Theory of Knots and Braids. An Introduction. North–Holland, Amsterdam, 1983.

[11] R.S. Pierce: Associative Algebras. Graduate Texts in Mathematics 88, Springer–Verlag, 1982.

[12] E.H. Spanier: Algebraic Topology. McGraw–Hill, New York, 1966.

[13] N.E. Steenrod: The Topology of Fibre Bundles. Princeton Univ. Press, 1951.

[14] B.L. van der Waerden: Algebra I. Grundlehren der Mathematischen Wissenschaften 33, Springer–Verlag, Fünfte auflage 1960.

[15] G.W. Whitehead: Elements of homotopy theory. Graduate Texts in Mathematics 61, Springer–Verlag, 1978.

## Surveys

[16] V.L. Hansen: Polynomial covering maps. In: Braids (eds. J.S. Birman and A. Libgober). Contemporary Mathematics 78, Amer. Math. Soc., 1988, 229–243.

[17] V. Ja. Lin: Artin braids and the groups and spaces connected with them. Itogi Nauki i Tekhniki, Algebra, Topologiya, Geometriya 17(1979), 159–227, and J. Soviet Math. 18(1982), 736–788.

[18] W. Magnus: Braid Group: A survey. In: Proceedings of the Second International Conference on The Theory of Groups. Lecture Notes in Mathematics 372, Springer–Verlag, 1974, 463–487.

## Papers

[19] J.W. Alexander: A lemma on systems of knotted curves. Proc. Nat. Acad. Science USA 9(1923), 93–95.

[20] J.W. Alexander: Deformations of an n–cell. Proc. Nat. Acad. Science USA 9(1923), 406–407.

[21] R. Arens and K. Hoffman: Algebraic extensions of normed algebras. Proc. Amer. Math. Soc. 7(1956), 203–210.

[22] E. Artin: Theorie der Zöpfe. Abh. Math. Sem. Univ. Hamburg 4(1925), 47–72.

[23] E. Artin: Theory of braids. Ann. of Math. (2) 48(1947), 101–126.

[24] J.S. Birman: On braid groups. Comm. Pure and App. Math. 22(1969), 41–72.

[25] A Dold: Partitions of unity in the theory of fibrations. Ann. of Math. (2) 78(1963), 223–255.

[26] T. Duchamp and R.M. Hain: Primitive elements in rings of holomorphic functions. J. Reine Angew. Math. 346(1984), 199–220.

[27] J.L. Dyer: The algebraic braid groups are torsion–free: an algebraic proof. Math. Zeitschrift 172(1980), 157–160.

[28] E. Fadell: Homotopy groups of configuration spaces and the string problem of Dirac. Duke Math. J. 29(1962), 231–242.

[29] E. Fadell and J. van Buskirk: The braid groups of $E^2$ and $S^2$. Duke Math. J. 29(1962), 243–258.

[30] E. Fadell and L. Neuwirth: Configuration spaces. Math. Scand. 10(1962), 111–118.

[31] R.H. Fox and L. Neuwirth: The braid groups. Math. Scand. 10(1962), 119–126.

[32] E.A. Gorin and V. Ja. Lin: Algebraic equations with continuous coefficients and certain questions of the algebraic theory of braids. Mat. Sb. 78(120) (1969), 579–610, and Math. USSR Sbornik 7(1969), 569–596.

[33] V.L. Hansen: Coverings defined by Weierstrass polynomials. J. Reine Angew. Math. 314(1980), 29–39.

[34] V.L. Hansen: Polynomial covering spaces and homomorphisms into the braid groups. Pacific J. Math. 81(1979), 399–410.

[35] V.L. Hansen: Algebra and topology of Weierstrass polynomials. Expositiones Mathematicae 5(1987), 267–274.

[36] V.L. Hansen: A model for embedding finite coverings defined by principal bundles into bundles of manifolds. Topology and its Applications 28(1988), 1–9.

[37] V.L. Hansen: The characteristic algebra of a polynomial covering map. Math. Scand. (to appear).

[38] A.A. Markov: Über die freie Äquivalenz der geschlossenen Zöpfe. Recueil Mathematique Moscou = Math. Sb. 1(1935), 73–78.

[39] J.M. Møller: On polynomial coverings and their classification. Math. Scand. 47(1980), 116–122.

[40] M.H.A. Newman: On a string problem of Dirac. J. London Math. Soc. 17(1942), 173–178.

[41] P. Petersen, V: Fatness of covers. J. Reine Angew. Math. (to appear).

# INDEX

Printed in the United States
By Bookmasters